Situational
Awareness

The McGraw-Hill *CONTROLLING PILOT ERROR* Series

Weather
Terry T. Lankford

Communications
Paul E. Illman

Automation
Vladimir Risukhin

Controlled Flight into Terrain (CFIT/CFTT)
Daryl R. Smith

Training and Instruction
David A. Frazier

Checklists and Compliance
Thomas P. Turner

Maintenance and Mechanics
Larry Reithmaier

Situational Awareness
Paul A. Craig

Fatigue
James C. Miller

Culture, Environment, and CRM
Tony Kern

CONTROLLING PILOT ERROR

Situational Awareness

Paul A. Craig

McGraw-Hill

New York   Chicago   San Francisco   Lisbon   London   Madrid
Mexico City   Milan   New Delhi   San Juan   Seoul
Singapore   Sydney   Toronto

Cataloging-in-Publication Data is on file with the Library of Congress

McGraw-Hill

*A Division of The **McGraw·Hill** Companies*

Copyright © 2001 by The McGraw-Hill Companies, Inc. All rights reserved.
Printed in the United States of America. Except as permitted under the United
States Copyright Act of 1976, no part of this publication may be reproduced or
distributed in any form or by any means, or stored in a data base or retrieval
system, without the prior written permission of the publisher.

1 2 3 4 5 6 7 8 9 0   DOC/DOC   0 7 6 5 4 3 2 1

ISBN 0-07-137321-7

*The sponsoring editor for this book was Shelley Ingram Carr, the editing supervi-
sor was Steven Melvin, and the production supervisor was Pamela Pelton. It was
set in Garamond per the TAB3A design by Deirdre Sheean and Paul Scozzari of
McGraw-Hill's Hightstown, N.J., Professional Book Group composition unit.*

Printed and bound by R. R. Donnelley & Sons Company.

 This book is printed on recycled, acid-free paper containing a
minimum of 50% recycled de-inked fiber.

McGraw-Hill books are available at special quantity discounts to use as premi-
ums and sales promotions, or for use in corporate training programs. For more
information, please write to the Director of Special Sales, Professional Pub-
lishing, McGraw-Hill, Two Penn Plaza, New York, NY 10121-2298. Or contact
your local bookstore.

Contents

Series Introduction

The Human Condition

The Roman philosopher Cicero may have been the first to record the much-quoted phrase "to err is human." Since that time, for nearly 2000 years, the malady of human error has played out in triumph and tragedy. It has been the subject of countless doctoral dissertations, books, and, more recently, television documentaries such as "History's Greatest Military Blunders." Aviation is not exempt from this scrutiny, as evidenced by the excellent Learning Channel documentary "Blame the Pilot" or the NOVA special "Why Planes Crash," featuring John Nance. Indeed, error is so prevalent throughout history that our flaws have become associated with our very being, hence the phrase *the human condition*.

The Purpose of This Series

Simply stated, the purpose of the Controlling Pilot Error series is to address the so-called human condition, improve performance in aviation, and, in so doing, save a few lives. It is not our intent to rehash the work of over a millennia of expert and amateur opinions but rather to *apply* some of the more important and insightful theoretical perspectives to the life and death arena of manned flight. To the best of my knowledge, no effort of this magnitude has ever been attempted in aviation, or anywhere else for that matter. What follows is an extraordinary combination of why, what, and how to avoid and control error in aviation.

Because most pilots are practical people at heart— many of whom like to spin a yarn over a cold lager—we will apply this wisdom to the daily flight environment, using a case study approach. The vast majority of the case studies you will read are taken directly from aviators who have made mistakes (or have been victimized by the mistakes of others) and survived to tell about it. Further to their credit, they have reported these events via the anonymous Aviation Safety Reporting System (ASRS), an outstanding program that provides a wealth of extremely useful and *usable* data to those who seek to make the skies a safer place.

A Brief Word about the ASRS

The ASRS was established in 1975 under a Memorandum of Agreement between the Federal Aviation Administration (FAA) and the National Aeronautics and Space Administration (ASRS). According to the official ASRS web site, *http://asrs.arc.asrs.gov*

> The ASRS collects, analyzes, and responds to voluntarily submitted aviation safety incident reports in order to lessen the likelihood of aviation accidents. ASRS data are used to:
>
> • Identify deficiencies and discrepancies in the National Aviation System (NAS) so that these can be remedied by appropriate authorities.
>
> • Support policy formulation and planning for, and improvements to, the NAS.

- Strengthen the foundation of aviation human factors safety research. This is particularly important since it is generally conceded *that over two-thirds of all aviation accidents and incidents have their roots in human performance errors* (emphasis added).

Certain types of analyses have already been done to the ASRS data to produce "data sets," or prepackaged groups of reports that have been screened "for the relevance to the topic description" (ASRS web site). These data sets serve as the foundation of our Controlling Pilot Error project. The data come *from* practitioners and are *for* practitioners.

The Great Debate

The title for this series was selected after much discussion and considerable debate. This is because many aviation professionals disagree about what should be done about the problem of pilot error. The debate is basically three sided. On one side are those who say we should seek any and all available means to *eliminate* human error from the cockpit. This effort takes on two forms. The first approach, backed by considerable capitalistic enthusiasm, is to automate human error out of the system. Literally billions of dollars are spent on so-called human-aiding technologies, high-tech systems such as the Ground Proximity Warning System (GPWS) and the Traffic Alert and Collision Avoidance System (TCAS). Although these systems have undoubtedly made the skies safer, some argue that they have made the pilot more complacent and dependent on the automation, creating an entirely new set of pilot errors. Already the automation enthusiasts are seeking robotic answers for this new challenge. Not surprisingly, many pilot trainers see the problem from a slightly different angle.

Another branch on the "eliminate error" side of the debate argues for higher training and education standards, more accountability, and better screening. This group (of which I count myself a member) argues that some industries (but not yet ours) simply don't make serious errors, or at least the errors are so infrequent that they are statistically nonexistent. This group asks, "How many errors should we allow those who handle nuclear weapons or highly dangerous viruses like Ebola or anthrax?" The group cites research on high-reliability organizations (HROs) and believes that aviation needs to be molded into the HRO mentality. (For more on high-reliability organizations, see "Culture, Environment, and CRM" in this series.) As you might expect, many status quo aviators don't warm quickly to these ideas for more education, training, and accountability—and point to their excellent safety records to say such efforts are not needed. They recommend a different approach, one where no one is really at fault.

On the far opposite side of the debate lie those who argue for "blameless cultures" and "error-tolerant systems." This group agrees with Cicero that "to err is human" and advocates "error-management," a concept that prepares pilots to recognize and "trap" error before it can build upon itself into a mishap chain of events. The group feels that training should be focused on primarily error mitigation rather than (or, in some cases, in addition to) error prevention.

Falling somewhere between these two extremes are two less-radical but still opposing ideas. The first approach is designed to prevent a reoccurring error. It goes something like this: "Pilot X did this or that and it led to a mishap, so don't do what Pilot X did." Regulators are particularly fond of this approach, and they attempt to regulate the last mishap out of future existence. These so-called rules written in blood provide the traditionalist with plenty of training materials and even come with ready-made case studies—the mishap that precipitated the rule.

Opponents to this "last mishap" philosophy argue for a more positive approach, one where we educate and train *toward* a complete set of known and valid competencies (positive behaviors) instead of seeking to eliminate negative behaviors. This group argues that the professional airmanship potential of the vast majority of our aviators is seldom approached—let alone realized. This was the subject of an earlier McGraw-Hill release, *Redefining Airmanship*.[1]

Who's Right? Who's Wrong? Who Cares?

It's not about *who's* right, but rather *what's* right. Taking the philosophy that there is value in all sides of a debate, the Controlling Pilot Error series is the first truly comprehensive approach to pilot error. By taking a unique "before-during-after" approach and using modern-era case studies, 10 authors—each an expert in the subject at hand—methodically attack the problem of pilot error from several angles. First, they focus on error prevention by taking a case study and showing how preemptive education and training, applied to planning and execution, could have avoided the error entirely. Second, the authors apply error management principles to the case study to show how a mistake could have been (or was) mitigated after it was made. Finally, the case study participants are treated to a thorough "debrief," where alternatives are discussed to prevent a reoccurrence of the error. By analyzing the conditions before, during, and after each case study, we hope to combine the best of all areas of the error-prevention debate.

A Word on Authors and Format

Topics and authors for this series were carefully analyzed and hand-picked. As mentioned earlier, the topics were taken from preculled data sets and selected for their relevance by ASRS-Ames scientists. The authors were chosen for their interest and expertise in the given topic area. Some are experienced authors and researchers, but, more important, *all* are highly experienced in the aviation field about which they are writing. In a word, they are practitioners and have "been there and done that" as it relates to their particular topic.

In many cases, the authors have chosen to expand on the ASRS reports with case studies from a variety of sources, including their own experience. Although Controlling Pilot Error is designed as a comprehensive series, the reader should not expect complete uniformity of format or analytical approach. Each author has brought his own unique style and strengths to bear on the problem at hand. For this reason, each volume in the series can be used as a stand-alone reference or as a part of a complete library of common pilot error materials.

Although there are nearly as many ways to view pilot error as there are to make them, all authors were familiarized with what I personally believe should be the industry standard for the analysis of human error in aviation. The Human Factors Analysis and Classification System (HFACS) builds upon the groundbreaking and seminal work of James Reason to identify and organize human error into distinct and extremely useful subcategories. Scott Shappell and Doug Wiegmann completed the picture of error and error resistance by identifying common fail points in organizations and individuals. The following overview of this outstanding guide[2] to understanding pilot error is adapted from a United States Navy mishap investigation presentation.

Simply writing off aviation mishaps to "aircrew error" is a simplistic, if not naive, approach to mishap causation. After all, it is well established that mishaps cannot be attributed to a single cause, or in most instances, even a single individual. Rather, accidents are the end result of a myriad of latent and active failures, only the last of which are the unsafe acts of the aircrew.

As described by Reason,[3] active failures are the actions or inactions of operators that are believed to cause the accident. Traditionally referred to as "pilot error," they are the last "unsafe acts" committed by aircrew, often with immediate and tragic consequences. For example, forgetting to lower the landing gear before touch down or hotdogging through a box canyon will yield relatively immediate, and potentially grave, consequences.

In contrast, latent failures are errors committed by individuals within the supervisory chain of command that effect the tragic sequence of events characteristic of an accident. For example, it is not difficult to understand how tasking aviators at the expense of quality crew rest, can lead to fatigue and ultimately errors (active failures) in the cockpit. Viewed from this perspective then, the unsafe acts of aircrew are the end result of a long chain of causes whose roots originate in other parts (often the upper echelons) of the organization. The problem is that these latent failures may lie dormant or undetected for hours, days, weeks, or longer until one day they bite the unsuspecting aircrew....

What makes the [Reason's] "Swiss Cheese" model particularly useful in any investigation of pilot error is that it forces investigators to address latent failures within the causal sequence of events as well. For instance, latent failures such as fatigue, complacency, illness, and the loss of situational awareness all effect performance but can be overlooked by investigators with even the best of intentions. These particular latent failures are described within the context of the "Swiss Cheese" model as preconditions for unsafe acts. Likewise, unsafe supervisory practices can pro-

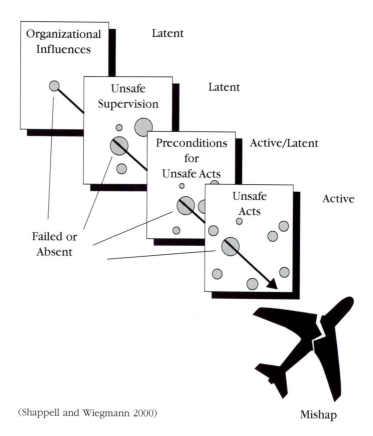

(Shappell and Wiegmann 2000)

mote unsafe conditions within operators and ultimately unsafe acts will occur. Regardless, whenever a mishap does occur, the crew naturally bears a great deal of the responsibility and must be held accountable. However, in many instances, the latent failures at the supervisory level were equally, if not more, responsible for the mishap. In a sense, the crew was set up for failure....

But the "Swiss Cheese" model doesn't stop at the supervisory levels either, the organization itself can impact performance at all levels. For instance, in times of fiscal austerity funding is often cut, and as a result, training and flight time is curtailed. Supervisors are therefore left with tasking "non-proficient" aviators with sometimes-complex missions. Not surprisingly, causal factors such as task saturation and the loss of situational awareness will begin to appear and consequently performance in the cockpit will suffer. As such, causal factors at all levels must be addressed if any mishap investigation and prevention system is going to work.[4]

The HFACS serves as a reference for error interpretation throughout this series, and we gratefully acknowledge the works of Drs. Reason, Shappell, and Wiegmann in this effort.

No Time to Lose
So let us begin a journey together toward greater knowledge, improved awareness, and safer skies. Pick up any volume in this series and begin the process of self-analysis that is required for significant personal or

organizational change. The complexity of the aviation environment demands a foundation of solid airmanship and a healthy, positive approach to combating pilot error. We believe this series will help you on this quest.

References

1. Kern, Tony, *Redefining Airmanship,* McGraw-Hill, New York, 1997.

2. Shappell, S. A., and Wiegmann, D. A., *The Human Factors Analysis and Classification System — HFACS,* DOT/FAA/AM-00/7, February 2000.

3. Reason, J. T., *Human Error,* Cambridge University Press, Cambridge, England, 1990.

4. U.S. Navy, *A Human Error Approach to Accident Investigation,* OPNAV 3750.6R, Appendix O, 2000.

Tony Kern

Foreword

Former Thunderbird lead and U.S. Air Force Chief of Staff (1990–1994) Merrill McPeak believed that one thing separated the good fighter pilot from the great one—situational awareness, or SA for short. McPeak believed that if a pilot were able to build and keep an accurate mental picture of the near-term past and present, he would be successful in creating a positive outcome for the future.

Far too often, we see the downside of this equation play out tragically; a pilot who loses altitude and terrain awareness, distracted by a confusing approach plate while descending, or an overaggressive military pilot who loses track of a formation member, only to end up in a midair collision with him. Good SA is a survival skill, and one that extends beyond the cockpit.

Situational awareness is a critical skill for each of us in our day-to-day lives. How often have we thought, "If I had only known _____, I would have done something different." Because SA extends into every corner of our lives, it becomes all the more important to us—and I strongly believe that the following pages will help you be a more effective and efficient person, as well as a safer pilot.

There is still a great deal to learn about how humans perceive events. Some of the finest minds in the world have worked for decades on the challenge of improving awareness of our surroundings. Some work under the premise that the mind can be augmented by technological devices that will improve our awareness through tactile, aural, and visual reminders. Stall-warning horns, stick shakers, TCAS, and GPWS are all examples of these efforts. Others try to eliminate the causes of lost SA through a variety of physiological and psychological efforts, such as fatigue mitigation and workload management. Still others look to a teamwork solution in an attempt to create "team SA," where each member of an aircrew works to improve the SA of the others and thereby enhance the group SA as a whole. Finally, there are human-factors experts who study breakdowns of situational awareness to increase our SA *about SA* (my head is starting to hurt just thinking about it).

Each of these initiatives has a place in improving effectiveness and safety, but for the average pilot to grasp the totality of the challenge, it needs to be pulled together in a single volume. Paul Craig has done the aviation community a huge favor with this effort. Using actual case studies from the ASRS, he operationalizes a sometimes-abstract concept. Using plain-language analysis, Paul cuts to the heart of three critical issues:

1. Maintaining situational awareness
2. Recognizing the loss of situational awareness
3. Recovering from lost situational awareness

Now this is a pretty nice toolkit all by itself, but the author goes much deeper, pointing out subtleties and differences between pilots and illuminating cues so that you can catch yourself before it is too late. As a flight surgeon of mine once told me, "Controlled flight into terrain (CFIT) is a humbling epitaph." This book will go a long way in helping each of us avoid that potential.

To cover this type of ground in a single volume takes an individual with deep roots in aviation theory and practice. Paul Craig has both in spades. As a practicing flight instructor and university professor, Craig combines the best of two worlds. He is the author of several highly acclaimed instructional books, including *How to Be a Better Pilot, Light Airplane Navigation Essentials, Multiengine Flying,* and most recently *Pilot in Command,* a compelling book that applies the term "command" to general aviation in a new and unique way.

I think you will find this volume to be both interesting and insightful. I know you will be a better pilot for having read it.

Fly smart.

Tony Kern

Situational Awareness

1

Fat, Dumb, and Happy

The first time I was introduced to the idea of situational awareness, it was done in a somewhat unsophisticated way. I had a college professor whose ways, terms, and philosophies were forged while flying in the Second World War. He often used the phrase, "fat, dumb, and happy" to describe a pilot who was dangerously unaware of his surroundings while in flight.

Of course, back then the whole science of situational awareness did not have a name. What we call situational awareness today was then an unspoken trait that experienced pilots just had and inexperienced pilots did not have. There was no formal training designed to acquire situational awareness; it was just a part of the normal seasoning process of veteran pilots. It you flew with situational awareness you were considered a pro; if you did not you were a pilot who was "fat, dumb, and happy."

The phrase "fat, dumb, and happy" was never further defined, but pondering it now, I see deeper meaning. Being fat, of course has nothing to do with the pilot's weight. The word fat is a metaphor for "well fed." Well fed means never having been hungry, never having to face any hardship. In this sense, "fat" means that the pilot has many advantages that can be quickly taken for granted. Today, pilots have safe airplanes to fly. We have excellent radio equipment available. We have color charts, and ground school videotapes, and satellites to guide us. The pilot has all the advantages in the world, but this easy condition can make the pilot complacent. When you use GPS to do your navigating every time, there will come a time when you can no longer plot the course on your own. Pilots get so used to a good thing that luxuries become essentials. They depend too heavily on the "extras," so when those extras stop working they are lost—unable to solve problems. They lose the ability to deal with the basics—they become fat.

Being "dumb" in this sense does not mean that the pilot is not book-smart. Pilots may know an airplane's every V speed, yet still be unaware of what is taking place around them. A pilot described as dumb in this context means that the pilot is not seeing clues to airplane problems as they arise. The problems are mounting and compounding, but the pilot is not yet aware of the problems. If an airplane's oil pressure gauge slowly begins to indicate a loss of pressure, the pilot should spot this, verify the problem with other instruments, and start looking for a place to land. At the very least the pilot should remain over flat terrain. But if the pilot does not see the creeping gauge, the problem will get worse and worse while the pilot takes no corrective actions to reduce the risks of an engine failure. The unaware pilot first becomes aware when the engine quits. But now the pilot is over wooded and hilly terrain with no good landing options. The opportunity to mitigate the situation was lost because the pilot did not know a dangerous situation was even present. The pilot was caught off guard because he or she was dumb to the situation.

Being "happy" as defined in the phrase means more than just being in a good mood. Being happy goes beyond just being unaware. Being in this state of "happiness" means that the pilot is in danger without knowing it, and at the same time the pilot does not know how to detect the danger. If the oil pressure gauge begins to indicate a loss of pressure, the "happy" pilots might see the gauge, but are unable to put two and two together. They see the needle on the red line, but it does not enter their minds that this is something that they should be concerned with. They figure that the gauge must be wrong because nothing bad could happen to them. The image I get is of people gleefully skipping through a mine field. There is danger all around, but they are oblivious to it and are just too busy having a good time. They are happy. Sometimes pilots will enter into this happy state because they carry a large dose of overconfidence and even arrogance.

Taken all together, the phrase fat, dumb, and happy paints the picture of pilots who have been sheltered, become complacent, fly with dulled senses, and allow themselves to progress into ever-increasing dangers before it finally dawns on them what is happening, but by then it is too late. Almost every accident scenario will contain all or part of the fat, dumb, and happy concepts. When accidents are avoided, it usually is because pilots were not fat, dumb, or happy and they preempted the danger with their awareness.

The Aviation Safety Reporting System (ASRS), known more commonly to pilots as the "NASA form" program, provides the story of a flight crew who were faced with a problem that challenged their ability to go back to basics. The story is told by the first officer of an L1011-500 airliner.

ASRS NUMBER 368245 It was a routine transoceanic flight westbound from Madrid to Atlanta, Georgia. At about 30 degrees west, an electrical failure occurred which put red flags into view on most of the captain's flight instruments, engine instruments, and fail lights on the overhead panel. It was not, however, a total electric failure. We all discovered the event at the same time. There were no other warnings or chain of events leading up to this point. Upon further investigation, we discovered that the line powering the essential aircraft bus had failed and that the "automatic" switching to an alternate source had not occurred. This was the crux of the confusion. For a few moments we felt that "this can't be happening to me" scenario, but we eventually came to grips with the fact that the essential bus was causing all our problems. We switched to another source of power manually and things went back to normal for a while! Shortly thereafter, the aircraft essential bus dumped again. We elected to continue status quo with the first officer as the pilot flying the airplane. We decided to reach Canada before fooling with any more alternatives. We also debated the merits of continuing on to our final destination versus landing short. We arrived at the conclusion that with good flying conditions, with excellent destination weather, we could navigate on what we had and there would be no risk. Lessons learned: Don't trust automatic backups and use good cockpit resource management.

The crewmembers, without warning, lost all their computerized cockpit displays, but they were not so reliant on those displays that they could not find Atlanta, Georgia, without them. They used the high-tech equipment, but they were also ready and able to use low-tech equipment in a pinch. They had not become complacent in the basics, and the passengers never knew the difference.

Unfortunately, a flight instructor and student flying in a Cessna 172 were caught by complacency on a flight when they inadvertently penetrated a Class C airspace without a clearance. This story is told by the flight instructor:

ASRS NUMBER 426980   While training in the area of Hastings, Michigan, I feel that I may have come close to entering the Grand Rapids (Kent county) airspace without a clearance to do so. We had no warning on GPS. But after figuring out our position, I realized how close we had gotten. The wind must have blown us further north than I expected. When I realized where we were, we took up a heading of south to clear the airspace.

The instructor had in fact flown into the Class C airspace. The instructor had been relying on a handheld GPS unit to tell him if he was getting too close, but the GPS provided no warning. At some point the airplane flew into the airspace without authorization, and in that instant the instructor was unaware of the problem. He had placed his trust in an instrument, but the instrument failed to perform as expected, leaving the pilot unprotected and unaware of his own situation. He had become fat with complacency.

Accidents have taken place simply because pilots did not recognize a problem as it occurred. Take the example of this student pilot who became confused about an airport elevation:

NTSB NUMBER LAX96LA048   The airport elevation at Imperial County Airport is 56 feet below sea level. In the pilot's navigation logs under the destination airport field elevation column, "-56" was entered with "950" written next to it. The traffic pattern altitude is listed as "944 MSL (1,000 AGL)" in the Airport/Facility Directory. The FBO line person who communicated with the pilot observed the pilot's first landing approach and reported that it appeared to be a little high. The aircraft (a Cessna 150) then initiated a go-around and made right traffic for runway 32. Ground witnesses reported that during the aircraft's traffic pattern following the go-around, the aircraft flew an extremely low pattern estimated at only 100 feet AGL. The pilot then contacted the FBO line person on the radio a second time. The witness stated that the pilot seemed confused about the traffic pattern altitude and airport elevation. She repeatedly asked what the field elevation was, and the witness responded with 60 feet below sea level. The witness stated that the pilot responded, "I thought it was about 900 feet." The pilot then terminated the second approach with another go-around. According to the witness the third traffic pattern was flown about the same AGL altitude as the second one. Shortly after the pilot reported turning base, the power went off at the airport. A ground witness near the accident site saw the aircraft turn at a very low altitude and collide with a 40-foot tall power pole about ½ mile southeast of the airport. Detailed examination of the aircraft revealed no discrepancies in any system. Review of the sun's location revealed that as the pilot turned from downwind to base, she would have been looking almost directly into the sun.

The NTSB later concluded that the probable cause of the accident was, "The student pilot's misreading of the altimeter in the traffic pattern, which put the aircraft in dangerous proximity to the ground and obstructions. A factor in the accident was the pilot's inability to see the power pole due to sun glare as the aircraft turned onto base from the downwind leg."

Tragically, the student pilot was killed in the accident. The altimeter does not have a "negative" range. When the needle passes below sea level, an indication of −50 could easily be seen as +950. There was even a notation of this fact on the recovered navigation log, but apparently this pilot made this error anyway. She thought she was 1000 feet in the air when in fact she was near ground level. But this misinterpretation of the instrument could have been detected by looking out the window and realizing that the downwind altitude was much closer to the ground than it should have looked. She must have dealt with this conflict, because she radioed to ask again about the airport elevation. The conflict was between what she saw out the window and what she saw on the instrument. She was not able to reconcile the two before it was too late. There were clues to the situation that this pilot did not see or did not use to resolve the problem.

An even more blatant case of unawareness was reported to NASA by this pilot of a Cessna 152.

ASRS REPORT NUMBER 409766 As I neared my destination and shortly after being cleared to 3,000 feet, my engine stopped. My assessment was lack of fuel. I was told that the aircraft had 4 hours of usable fuel and was basing my decisions on that. I now believe that the time estimate was too generous. I landed the aircraft in a stubble field away from people and property and had declared an emergency. The Hobbs meter indicated 3.4 hours of flight. The landing was made safely without damage to property or the airplane. There were no injuries. The airplane was inspected and then flown off the highway nearby. Police were there to assist and hold back traffic.

This pilot had flown for over 3 hours without doing any fuel calculations, other than what someone "told" him was 4 hours of fuel. During every minute of that flight a situation was developing that the pilot was completely unaware of. It only first dawned on this pilot that he had been "dumb" when the engine stopped. He was extremely fortunate that there happened to be a safe place to land available when the fuel ran out—it was not because of good planning that this pilot survived.

Survival is not assured when other factors are involved. Factors such as a dark night and a pilot who is not as familiar with his airplane as he should be came together in the following example.

NTSB REPORT NUMBER FTW89FA155 The aircraft had been refueled at a nearby airport before this pilot started his flight. He took off and flew the aircraft from Oklahoma City to Tishomingo, Oklahoma, with the fuel selector valve positioned to the right tank. The right tank was the tank that was used by the previous pilot. On the return flight to Oklahoma City at night, the pilot's father noticed the right fuel gauge indicated empty while the left gauge indicated full. The pilot then could not find the fuel selector valve. The pilot requested help from the Air Route Traffic Control Center. There happened to be a pilot at the ARTCC who was familiar with the PA-28 airplane that the pilot was flying, but by the time this pilot was located the airplane engine lost power. Information about the fuel selector position was passed to the pilot but it was too late. The airplane hit trees and crashed. Later the pilot said that he had never operated the landing gear or fuel selector during his airplane checkout. He thought the fuel selector was on the floor between the seats, but in fact it is located on the left wall of the cockpit near the pilot's left knee. The fuel selector was never moved and was found positioned toward the empty right tank.

The NTSB probable cause:

Improper planning/decision making by the pilot, his lack of understanding of the procedures for operating the airplane, his improper use of the fuel selector in managing the fuel supply, fuel starvation, and the pilot's failure to know and follow the emergency procedures when the engine lost power. Contributing factors were: Inadequate transition training provided by the instructor pilot, the pilot's lack of familiarity with the airplane, the dark night, and the trees.

Think back to this pilot prior to when the trouble began. He was unaware of the fuel situation. He was unaware of how the airplane's systems worked. He was not monitoring his fuel consumption. He was flying an airplane that was in perfect working order, and he just expected it to remain that way. When a problem was first detected it was not the pilot, but a passenger who brought the fuel level to the attention of the pilot. This pilot certainly fits the concept of fat, dumb, and happy. But the terrible tragedy of the story is that the passenger, who first saw the problem and was the pilot's father, was killed in the accident.

When you read about some pilots who got into trouble, you just wish you could have been there to shake them back into awareness. They go deeper and deeper into danger and cannot see what is taking place until it is too late. The following three examples all involve pilots who could not or would not wake to reality.

NTSB REPORT NUMBER SEA95FA031   Local adverse weather conditions, including low ceilings, and snow were reported moving west to east as the airplane flew east to west. The noninstrument rated private pilot received a weather briefing for a VFR flight over 5 hours before he actually departed. He and his three passengers departed at night, in mountainous terrain and in VFR conditions with the intention of flying to an airport located 90 miles away for dinner. The pilot received ATC radar advisories and reported that the ceilings were getting lower along his route of flight. He was advised by ATC that areas of level one and two precipitation existed in front of him. The airplane continued to descend after ATC services were terminated. Radar data for the airplane was lost shortly thereafter. The airplane impacted a mountain ridge about 6,200 feet MSL and was destroyed. The ridge is located along a direct line from the departure airport to the destination airport. No distress calls were recorded from the pilot, and no evidence of preimpact mechanical deficiencies were found. The probable cause statement for the accident read, "The VFR pilot's attempt to continue the flight into instrument meteorological conditions, and his failure to maintain clearance with the mountainous terrain."

NTSB REPORT NUMBER ANC97LA025 The noninstrument rated private pilot and two passengers were on a cross-country flight on top of an overcast layer of clouds at 10,000 feet MSL when the pilot radioed FAA air traffic controllers for assistance. The pilot told controllers he thought he was just a few miles from Anchorage, Alaska, his intended destination, but he was actually about 124 miles northwest of Anchorage. The pilot was asked if he could turn towards and cross a nearby mountain range to reach VFR conditions. The pilot indicated he did not have enough fuel left, and the he was presently flying through the tops of the overcast. During his communications with controllers the pilot noted a marked disparity between his wet compass and his gyro-driven heading indicator; he also said his only electronic navigation instrument on board, a loran, was not reliable. Radio contact was lost with the pilot, and soon thereafter, an Emergency Locator Transmitter was heard. The airplane was discovered crashed in a near vertical position on a glacier. Postaccident inspection disclosed no mechanical anomalies with the airplane and a functional loran. About five to six gallons of fuel was remaining in the left wing fuel tank.

Probable Cause: "The pilot's continued VFR flight into instrument meteorological conditions, and subsequent failure to maintain control of the airplane. Factors associated with the accident are the pilot's inadequate weather evaluation, his becoming lost/disoriented, and spatial disorientation."

NTSB REPORT NUMBER ATL84FA095 Prior to departure, the pilot was advised twice that his destination weather was below minimums. The noninstrument rated pilot departed Dulles Airport on an IFR flight plan into instrument conditions. After departure, he did not follow assigned headings and for a period of about 30 minutes, radio communications were lost. After radio contact was reestablished, the pilot was continuously advised that the weather at his destination (Chesapeake, Virginia) was below approach minimums. The pilot subsequently stated that he wanted to put down at the first available airport. Again radio contact was lost until the pilot reported he was flying inland toward Tri-County Airport in Ahoski, North Carolina. This was the last transmission that was received. The pilot changed his transponder code to 1200. The aircraft impacted the ground 1 and 1/2 miles south of the pilot's original destination airport.

All of the last three accidents were fatal accidents. Each involved a pilot who early on thought that everything was okay. Each pilot continued to believe that all was well until the situation progressed to a point of no return. The pilots were in trouble but did not know they were in trouble. They happily proceeded to the site of the accident.

Then sometimes an accident takes place that gives even a different meaning to the concept of "happy." NTSB accident investigation number ATL91LA088 is such a case:

The 24-year-old pilot flew his airplane at a low altitude over the surface of the Atlantic Ocean near Kure Beach, North Carolina. A witness reported that the airplane's landing gear hit a wave, and then the airplane nosed over. The pilot was rescued, but the only other passenger drowned. Toxicology testing of the pilot after the accident revealed a blood/alcohol level of 0.165%. The pilot had flown the airplane without the owner's consent. The wreckage was not recovered from the ocean.

The FAA regulations require an eight-hour period between the consumption of alcohol and the operation of an airplane. In addition, the FAA prohibits the operation of any aircraft when the blood/alcohol level is above 0.04%, no matter how long the pilot waited after consuming alcohol. This pilot was legally considered to be under the influence of alcohol when he stole the airplane, crashed it into the ocean, and killed his passenger. One of the effects of alcohol consumption is euphoria, which is an exaggerated sense of well being—or happiness. It is hard enough for pilots to maintain awareness without having a chemical in the bloodstream to make it worse.

The goal of pilots today is to become "skinny, smart, and suspicious." In other words, we need pilots who can use new technology without giving up the basics, who can spot trouble in advance and troubleshoot to preempt a crisis, and who are wary, savvy, and defensive, not completely trusting anything, but able to use all available resources to make safe decisions. This kind of pilot would display and use situational awareness. The rest of this volume is an adventure in gaining and maintaining situational awareness as you fly so that you become the best and safest pilot possible.

2

Discovering Situational Awareness

General aviation is often the last to be invited to the party. When computers and flight simulators were married to form a virtual flight training environment, general aviation pilots were priced out of the market. When cockpit resource management came of age, the emphasis was on military and airline crews. Now situational awareness is being taught to merge all that has been previously learned, but these lessons learned only trickle down to the general-aviation pilot. We need a new way to use situational awareness techniques that brings them all the way down to the pilot of a Cessna 150.

Vince Mancuso, coordinator of line operational simulation for Delta Airlines, says, "The most advanced work in situational awareness has been done in the military." The term situational awareness (SA) was coined in the military, and it is easy to understand why. There can be no more intense pilot situation than air combat. The pilot is called upon to monitor the airplane and its systems, the airplane's weapons, aerobatic maneuvers, and, of course, the enemy all at once, with death being a result of failure. The ability to control all these dynamic forces successfully was called the "Ace Factor" in a 1989 book by Mike Spick. He reports that the top scoring aces of World War II generally avoided high-confusion melees and excelled by picking off stragglers. He said, "The aces were very aware of their own limitations and tended to keep out of situations with which they could not cope adequately." Fighter pilots have often been faced with situations where having situational awareness or not meant returning home or not. It became a clear-cut SA/death choice. This generated a great need to gain, keep, and manage situational awareness. So the military began to examine what it meant to have situational awareness and designed curriculum aimed to teach situational awareness techniques.

Then aviation changed. The airlines grew and flight "crews" dominated the peacetime sky. The single-seat flight pilot who had become known as a romantic hero became out of place on an airliner's flight deck. In his book *The Right Stuff*, Tom Wolfe portrayed the fighter pilot as a lone warrior who would fly to the edge of eternity and back single-handedly. This pilot was independent, hard nosed, had complete authority, and communicated only as a last resort. Dr. H. C. Foushee, working for ASRS, concluded that this solitary fighter pilot approach was necessary in battle, but dangerous when a "team" environment was needed. Today on the flight deck of airliners, groups of two and three crewmembers must share the workload. In this world, communication and coordination are required. You can see that fighter pilots from a closed environment would have problems transitioning to an open, shared-responsibility environment. Some could not make the change. Some made the change but not smoothly. The need to maintain awareness was the same, but how to achieve it inside a group was different.

So researchers went back to work, this time observing crews, not just single pilots. Taking advantage of the new generation of flight simulators, soon there was a mountain of research. But sometimes research has a hard time making it from the laboratory to the cockpit. "Research reports aren't written in a way that an airline manager or pilot can use," says Mancuso. "They require a lot of work and translation." To attack the problem Mancuso helped form the CRM developers group, "I knew there had to be a way to build a bridge between all the great research that's going on and the people who are designing the pilot training programs, so ultimately the pilot would be the recipient."

The bridge was built, but it was built to span the gap between the research and the airlines and the military. Any benefit to general aviation has again been whatever has trickled down.

The research has even received attention outside aviation. Situational awareness and crew recourse management techniques are now being employed in the medical profession. It turns out that the communication and coordination skills that are needed between pilots who are members of a crew and between single pilots and air traffic controllers are the same skills that emergency and surgical teams need. In 1993 a group of researchers analyzed 2000 reports on critical medical incidents. The conclusion was that 70 to 80% of medical mishaps are caused by human factors. That is approximately the same as the percentage of aircraft accidents that are attributed to pilot error. Later, additional medical research was conducted when observers were present during 90 surgical operations. The observers were particularly interested in the communication and coordination between the surgical and anesthesia teams. As you might imagine the observers saw many examples of poor coordination, including, "failure to communicate insufficient regional anesthesia prior to incision." Ouch! The report went by the name, "Jumpseating in the Operating Room," which was a clear association with lessons learned observing from an airliner's jumpseat.

A Harvard University study has estimated that there are 98,000 deaths each year due to "medical errors." This means that, depending on the definition used for "medical error," your doctor is about 900 times more likely to kill you than your pilot.

Medical error and pilot error are similar in that both can be reduced with better awareness. The medical profession has begun to recognize the benefits of the aviation-based research. Just like pilots who train in a flight simulator, now at some medical schools physicians in training use a "patient simulator." A patient simulator is a life-size human body mock-up. The patient simulator breathes, has a temperature, a heart beat, and through a computer can be made to have particular symptoms. Flight crews use "line-oriented flight training" (LOFT) to create real-world scenarios in a flight simulator. Physicians training on the patient simulator likewise face real-world scenarios. What happens when a patient goes into cardiac arrest or has an allergic reaction to medication? Will the physicians solve the problem while working as a team, or will they fall all over themselves while arguing over the best course of action? Do the physicians communicate effectively or is the lead physician unreceptive to information and suggestions? These are the same awareness challenges faced by pilots in flight. Pilots and doctors must both maintain situational awareness to be effective and safe.

It is true that the awareness and CRM research have been geared for the military and the airlines, but general aviation has received some attention. There has been some direct benefit to general aviation if you look. Constance Bovier, writing for *Flying Careers,* says, "From the first lesson, pilots learn that swift reactions are good...but not good enough. Getting ahead of the airplane is better. Staying ahead of the airplane is best. [This is] Situation Awareness 101."

Now let's take the final step. Let's build the bridge all the way to your cockpit. What does situational awareness mean to you when you fly?

There is a continuous loop that forms between information that comes in to you while you fly and the decisions you make. Most of these happen without your giving it much thought. There you are on final approach and you notice that you are somewhat high. You instinctively make a power adjustment and/or lower another notch of flaps. The problem is preempted before it ever gets started. Because you were aware that a problem was developing, you applied corrective action, solved the problem, and kept on going. If you had not become aware of the airplane getting too high, the problem would have persisted and gotten worse. You might be forced to make a go-around, but your awareness saved the day.

Being too high on final approach and recognizing it in time is, of course, a simple example. The flight environment can be very complex, with several evolving problems taking place simultaneously. The pilot must achieve and maintain awareness of everything, but how is this accomplished? Balance. The pilot must strike a balance between "the big picture" and "the small details." Situational awareness is really another way of describing focused attention. Imagine that you focus your attention through the lens of a camera. Some lenses are "wide-angle" and bring in the big picture.

Other lenses can "zoom in" to see a single spot within the big picture. To achieve situational awareness, you must know when to use the correct lens.

In most situations, the widest possible view is needed. Photographers often use a "fish-eye" lens that can bring almost 180 degrees into the frame. This extremely wide angle will distort the periphery, but includes the greatest amount of information. Pilots should start out with their attention focused through the wide-angle lens, but be ready to switch to the zoom. When the pilot observed that he was too high on final approach, he had to quickly focus attention on that problem. A zoom lens was temporarily needed to adjust the power and flaps. In that moment, other items that were in the big picture (traffic, radio communications, wind drift) were unseen because the focus was narrowed to that one item. As soon as the problem was solved, the pilot again took the wide-angle view and again became aware of other factors in the big picture. The trick is to balance the use of the wide-angle and zoom lens. Using the wide-angle all the time means seeing too much and missing details. Using the zoom all the time means fixing on a detail and ignoring everything else.

When two pilots fly together, they should never be using the same lens. If a problem comes up, one pilot should zoom in on the problem, while the other keeps track of the big picture. That is good advice for crews, but when you fly without another pilot you can't delegate; you must switch lenses appropriately. The proper balance requires the pilot to prioritize. There will be times when you must zoom in on a particular item, but you cannot stay zoomed in long because every second that you are using the zoom lens, something else may have crept into the periphery of the big picture. You can only work on one problem so long before you must leave it, check the big picture, and then come back.

Sometimes we must control the input of information into the big picture in order to buy the time necessary to work on the detail. What should you do if you enter the downwind leg of a traffic pattern, place the landing gear handle down, but the gear does not come down? You should first exit the traffic pattern. Get away from the crowd, and get some altitude. This will eliminate many of the factors in the big picture that you would ordinarily have to deal with and will allow you to zoom in on the problem. It is very hard to pump down the gear and fly the traffic pattern. That would require focusing attention both big and small. You cannot safely change lenses that fast, so reduce the need for the wide-angle lens so that you can use the zoom or even a microscope for a short time.

Then there are times when you have no control over time and the pilot must rapidly switch attention from big to small to big. When the pilot cannot keep up with the need to switch attention, an overload is sure to happen, and tragic results can follow. A pilot can reach a saturation point where the camera breaks altogether. Past that point the pilot is swept away by the events with no control and no awareness of what to do. Think of the pilot who flies into instrument conditions and becomes disoriented. There would be a swarm of information—some true, some false—flooding the pilot's brain. The pilot soon becomes overwhelmed, loses total awareness of the airplane's condition, and therefore cannot determine what to do. Many of these situations end when the pilot tears the airplane apart in the air in an attempt to regain awareness and control.

Sheryl Chappell wrote the seminal article "The Next Best Thing to a Crystal Ball" and compared the pilot's attention to aiming a flashlight. She wrote, "If you had a crystal ball, you would be aware of everything that is happening and is going to happen to your aircraft and the airspace you fly through, because the consequences of a lapse in awareness can be deadly."

Other times the pilot can misidentify a problem and aim the zoom lens in the wrong direction. Conducting a study with general-aviation pilots in a small flight training device, I observed many problem/solution mismatches. I gave over 50 pilots a scenario where engine roughness became apparent. While the engine ran rough, the oil pressure gauge slowly indicated a loss of oil. Meanwhile, the oil temperature gauge indicated a slow rise. Eventually, both gauges were indicating the red line. The pilots who focused their attention on these gauges recognized that an engine failure was imminent and that a quick landing was called for. But some of the pilots misdiagnosed the problem. They used their zoom lens on the carburetor heat valve. Thinking the problem was carburetor ice, they applied the heat and began believing that the problem was getting solved and that things were improving. They zoomed in on the wrong problem and their focus became too narrow to see the real problem.

So in order to reach any level of situational awareness, the pilot must skillfully use a narrow and a wide focus of attention. But it's not enough to be able to use the lenses; the pilot must know which lens is correct under different circumstances. The pilot must know how long to use one lens before switching to the other, and the pilot must know where best to aim the lens when it is used. It is a big job.

Successful in-flight decisions can only spring from being aware that a decision is necessary. Knowing a decision is necessary comes from being aware of the complete situation. Judith Orasanu, principal investigator for the ASRS Ames Research Center, says, "You can't have a decision out of an emergent situation in flight unless you're aware that you have a problem. That's where SA comes in."

When I was in college I went to see The Amazing Kreskin. Kreskin hypnotized fellow students, revealed secrets that only they could have known, appeared to read people's minds, and predicted things that later came true. One thing that Kreskin said that night has always stayed with me. He said that airline crews love it when they discover that Kreskin is on the flight. They figure that since he knows the future, he would not get on an airplane that was going to crash and so flying with Kreskin guarantees safety. Kreskin must have supernatural situational awareness.

But we do not have a crystal ball and we don't have Kreskin along on every flight, so we must learn to be aware on our own. Learning how to train your focus of attention leads to better awareness and better safety. Using case studies can aid in this training. The following are some examples taken from ASRS's Aviation Safety Reporting System where we can see the focus of other pilots. None of these examples represents a complete loss of situational awareness, but each does point to the fact that even a short period of confusion can lead to big problems. These reports are divided into three categories of awareness lapses. The first is a collection of forgotten clearances, headings, altitudes, or routes. The second group of reports consists of examples of pilots who became temporarily lost or misplaced. The third group consists of examples of pilots who became unaware of dangers that surrounded them.

Forgotten Clearances, Headings, Altitudes, and Routes

ASRS NUMBER 425954

A Flight Instructor takes off without an authorization at Melborne, Florida and has a near midair collision with a Challenger jet taking off on a crossing runway. The pilot tells the story:

Myself in the right front seat, a private student in the left front seat, and a private student in the rear seat, had finished our run-up and taxied to the hold short line of Runway 4 for takeoff. Our first call to the tower was interrupted twice by other aircraft on the frequency. It was quite a busy exchange for the controller, he did not seem able to talk fast enough before someone else would call in. When we did get ahold of him, we were told to hold short, which we did. While we waited, I was reviewing with the students the procedure for the soft field takeoff we would be doing. I heard the controller issue a clearance that sounded like the same clearance for takeoff that we had received so many times before. The student in the back seat was talking over the intercom at the time, so I asked the student controlling the aircraft if we were cleared for takeoff and he said yes. We proceeded with our takeoff roll and just before reaching takeoff speed, the student lost directional control. The aircraft started to skip off the left side of the runway with an extreme nose high attitude, without sufficient airspeed to climb. I had to wrestle with the student while also yelling to regain enough control to prevent a ground loop and/or hitting the runway lights. I managed to lower the nose enough to get the airspeed needed to climb. During the next few moments of the climbout, I would guess from about 50 feet to approximately 3001 feet, I did not hear anything from the tower. I think that there had been some radio traffic during our "emergency," but due to the urgency of my current situation and worried about our aircraft and ourselves, I don't remember hearing anything discernible but the stall warning horn and a jumble of different voices. At approximately 300–400 feet, the controller called us and asked if we realized that we just took off without a clearance and that we made a Challenger jet expedite his climb to miss us. He said to remain in the pattern and return to land and then call the tower. That is what we did next. I did not even see the jet until he had told us about it. At that point it was over 1 mile away. It had taken off Runway 9R and we were on Runway 4. They both meet at their departure ends. Looking back now at the situation, I can see how the busy training environment and all the radio calls were distracting for both me and the controller. The immediate "emergency" we had on the runway probably polarized me into dealing with that problem and shutting all other things out. (During our phone call the controller said that he tried to "cancel our clearance" when we "bobbled off the runway," but did not get a response.) Upon debrief with the students, the pilot swears he heard that he was cleared for takeoff. The back seat student said he saw the jet in the air but did not say anything until it had already passed overhead. (The controller during our phone conversation also mentioned hearing, "…cleared or clear (something)…," in the background on the tape, but said that he didn't say it.) A few of the other flight instructors on the frequency at the time, who came up to me afterwards to ask what happened, said they were confused when they heard the controller say that our "clearance was canceled," since he claims not to have given one. I guess unless I hear the tape of the event I won't know for sure. I do know that a little bit more attention to "the big picture" of what was happening around us could have prevented the entire situation.

This could have been much worse. The two airplanes could have collided at the end of their respective runways. Luck and the fact that a Challenger climbs faster than a Piper prevented a tragedy. But let's go back and see where the problem started. Let's look at where situational awareness was lost.

The pilot was waiting amid heavy radio traffic for a takeoff clearance. Ultimately, the instructor and the other two pilots aboard misunderstood their instructions. But what were the three pilots doing while they were waiting? "I was reviewing with the students the procedure for the soft-field takeoff we would be doing." Rather than listening and waiting, the instructor was creating a self-distraction. Any discussion of takeoff procedures should have been done before entering a "sterile cockpit" environment. The instructor changed lenses at an inappropriate moment. He should have been focusing on the big picture and said so in his closing statement, "I do know that a little bit more attention to "the big picture' of what was happening around us could have prevented the entire situation."

A "sterile cockpit" is the term used among flight crews that means those times when only essential flight-related conversation is allowed. General-aviation pilots should also adopt a sterile cockpit policy. Anytime while on the surface, in a traffic pattern, or within 10 miles of an airport, you should limit conversations to the specific problems at hand. The instructor of this near-miss said that "The student in the back seat was talking over the intercom at the time, so I asked the student controlling the aircraft if we were cleared for takeoff and he said yes." There was a conversation going on between the instructor and the person in the rear seat that blocked a controller transmission. When an intercom is in use and several people, pilots and nonpilots alike, are wired in, they should be told to remain quiet until advised otherwise by the pilot in command. This is required to maintain a sterile cockpit.

Finally, there was one more error that stands out. The instructor did not completely hear what the controller said, but instead of making sure he relied on an expectation. He said, "I heard the controller issue a clearance that sounded like the same clearance for takeoff that we had received so many times before." Since the instructor had heard the same clearance "so many times before," he assumed that the same clearance had been given again, even though he was not certain. If you are ever in doubt about any clearance or instruction, you must ask, not act.

ASRS NUMBER 425926

The pilot of a Cessna 414 was checking out another pilot in that airplane. During an instrument approach the airplane was allowed to descend dangerously low on the approach.

On an IFR flight we had an altitude excursion. We were assigned 2000 feet and were being given vectors on the ILS 23 to Morristown, New Jersey. As the captain on the flight, I was performing a checkout flight for a prospective pilot. This was my first flight with this pilot. It was his first time in the aircraft and the weather was solid IMC. As I studied the approach plate with my attention diverted from the instruments, I glanced up and noticed the altimeter read 1500 feet. I immediately began a recovery to 2000 feet. After the recovery was initiated, New York approach control called with a low altitude alert. We were stable at 2000 feet. This excursion could have been avoided by the pilot flying having a better instrument scan and also by better monitoring on my part as the pilot in command on this trip. Another mistake made on this flight was not departing prior to the void time given by Philadelphia approach control as we departed. We may have been as much as 2 minutes past our void time.

You can see the misaimed focus of attention throughout this example. The pilot in command temporarily used a narrow focus to examine an approach chart and in doing so lost the big picture, "As I studied the approach plate with my attention diverted from the instruments, I glanced up and noticed the altimeter read 1500 feet." The pilot flying the airplane also had a narrow focus and missed the fact that they were descending too soon. With both pilots on a narrow focus, neither was watching the big picture and soon both had lost awareness. With a new pilot checking out in a new airplane in instrument conditions, the need for prioritized focus was great. The pilot in command could have eliminated the need for such a high level of awareness by conducting the checkout when the weather was not "solid IMC." All the factors combined, added to a misaimed focus, lead to a CFTT—controlled flight toward terrain.

ASRS NUMBER 427829

A government flight crew in a Beech 18 deviated from an assigned route, entered the Los Angeles Class B airspace without authorization, and created a conflict with an airliner.

While flying the Hollywood Park route we flew inadvertently through the Los Angeles Class B airspace. We had a deviation off the 140 degree radial from the Van Nuys VOR. We were en route to a spray area over Woodland Hills and Chatsworth on the west end of the San Fernando Valley at 8500 feet. North of LAX we requested a heading change from the controller and was told to make that request of the next controller. When we checked on with the next controller, we were flying west of the 140 degree radial on an intercept for the first swath line. The request was made to the local controller to work in the spray region at 8500 feet. As I remember the region was clarified for the controller and then he called out an airliner, an MD80 at about our 10 o'clock position. We reported the MD80 in sight. We were then told to fly north to the Van Nuys VOR before proceeding any further west. We acknowledged and proceeded north until clear of the Bravo airspace. There were no

evasive maneuvers and separation from the MD80 is estimated to have been about 1500 feet laterally and 300–500
feet vertically. The MD80 passed well below and behind us.
The main contributing factor was my unfamiliarity with the
Hollywood Park route. I normally use the special flight rules
corridor where we were below the 5000 feet floor of the B
airspace. On the Hollywood Park route you are in the B airspace and should not deviate from the 140 degree radial
until clear. I became aware of the problem when the controller called the traffic and then instructed us to proceed
northbound. The problem could have been avoided with a
little more time spent in preparing for the route. Also it
could have been prevented if the controller had told me to
stay on the route until clear of the B airspace when I made
my request for a heading change. At that time I was handed
off to a new controller and assumed I was OK where I was
when in fact I was not. The fault was mine for the deviation
and I will take the appropriate measures to assure that it
does not occur again.

Here the focus was too big, so large that the pilot
missed a detail of the chart that was to be used for cir-
cumnavigating the Class B airspace. The pilot had the
big picture, but needed more attention to position.
Awareness could have been increased here with better
planning of the route. The pilot admitted that he was
unfamiliar with the route and said, "The problem could
have been avoided with a little more time spent in
preparing for the route."

ASRS NUMBER 427644

The pilot of a Cessna 210 accidentally follows the wrong
airway while talking to a flight watch about the weather.

First, let me say that this navigation error was my responsibility. The Error: Victor 23 reaches Clovis VOR Southeast bound
on a 124 degree heading and Victor 165 leaves the VOR on a
124 degree heading. However, Victor 23 (on which I was

filed) makes a jog to the right to 142 degrees. Unfortunately, the Victor 23 symbol is about 8 inches below on the airway. A quick glance makes one want to continue on a 124 degree heading (V165) which is what I did. However, distraction was the real culprit. Prior to reaching the VOR, I asked Fresno to go off frequency for a weather report. Permission was granted, and as we crossed the VOR I was in an intense conversation about the marginal weather ahead and at my destination with flight watch. By the time I got back on the center controller's frequency he barked, "Turn right 40 degrees, you are 5 miles from the airway centerline." I made the mistake of babbling some excuse and he repeated his command, which I executed. I was too embarrassed to say too much more and then he finally said (as if I was a new student) "You are now on the centerline." Definitely my fault—showing how creating a distraction for yourself right before a course or altitude change can screw you up.

A simple mistake was made because attention was diverted and fixation followed. Getting an in-flight weather update is a really good thing to do, but the pilot widened the focus of his attention to the big picture and missed the detail of the chart. Be careful not to inadvertently place yourself in a situation where wide and narrow focus will both be required. The pilot recognized later that he himself had set up the mistake, "Definitely my fault—showing how creating a distraction for yourself right before a course or altitude change can screw you up." It will happen enough when we have no control, so when you can avoid creating a situation that requires overlapping attention, do so.

ASRS REPORT NUMBER 427608

The pilot of a light twin engine airplane flies through a Class D airspace without authorization while en route to another Class D airport.

I called BDR tower 10 mi north for landing. I was told to report right base for Runway 24. I descended to pattern altitude (1500 feet for twins) and called tower with 2 miles right base. Tower said Sikorsky airspace (the adjoining Class D) was active and that I went through it. I have been based at BDR for several years and had never been aware of Sikorsky control zone being active on Sunday. Since I gave my position to the tower and they had me on radar, I believe they should have said Sikorsky was active. Published hours for Sikorsky are 1300Z to sunset Monday through Saturday. My flight was on a Sunday.

Whether the Sikorsky airspace was active and whether there might have been a NOTAM out on the airport's Class D status is not really the problem. The problem is that a pilot made an assumption based on an expectation. There are both good and bad habits. Flying without checking the most up-to-date information is a bad habit. This pilot flew into airspace where he had not been cleared. While flying in the unauthorized airspace he was completely unaware that anything was wrong. His attention focus should have started before the flight.

ASRS NUMBER 427275

A Cessna 150 and a DC9 had a near miss, 6 miles from Miami International airport. The Cessna pilot thought he was flying under the Class B airspace and the DC9 was on approach to Miami. Either the Cessna pilot accidentally penetrated the Class B or the DC9 was too low for that point along the approach. The pilot of the Cessna made the report:

My aircraft was level at 1000 feet MSL. Our position was 6 miles northwest of runway 9R at Miami International. We were within Class E airspace below Miami's Class B. Our heading was approximately south-southwest (190–195 degrees). We were in preparation for landing at TMB. A near miss occurred with DC9, or Fokker 100 type aircraft on approach to Miami's runway 9R. Distances provided on reverse side are rough estimates at best. No evasive action was immediately taken nor was it required. Miami's Class B airspace base begins at 1500 feet MSL in our position. The jet aircraft appeared to be on approach, below Class B airspace.

ASRS NUMBER 426919

In the next example, an air traffic controller does not follow specific procedures and as a result fails to hand off an airplane before it penetrates a Class B airspace. The controller lost situational awareness on the progress of the airplane and took a break before passing the airplane to the next controller.

Aircraft #1 was cleared from ILL to MSP to maintain 5000 feet through flight service. Aircraft #1 checked on the frequency departing the airport. I told him to report reaching 5000 feet. He never reported 5000 feet or turned on his transponder. He made it all the way to MSP airport before calling again. Mistakes I made were: One, I departed the aircraft in the computer but didn't write a time in Box 22 of the flight progress strip. Two, I did not manually start a track on aircraft. Three, I also took a break a few minutes after he departed and forgot to tell the relieving controller he had departed.

The controller is accepting responsibility, but the pilot never called to inquire about the approaching Class B airspace. Technically, since this was an IFR clearance all the way to MSP, the Class B was not violated, but the pilot must have known that he was getting close without getting passed on. One of the situational awareness traps is a situation when something that should have happened doesn't.

ASRS NUMBER 426900

The pilot of a small single-engine airplane flew to an airport without being aware that the airport had an operating control tower.

I did not realize the airport had a control tower. I attempted to contact UNICOM three times, entered the traffic pattern and made standard landing. I taxied to parking, and later in terminal I found out that the airport had a tower operating.

The pilot landed without a clearance at the airport inside a Class D airspace. There were no other traffic conflicts, but the pilot entered Class D without knowing it and landed, all the time being unaware of what was going on around him.

ASRS NUMBER 426707

A light twin-engine airplane with a student and flight instructor on board accidentally climbed into the underside of a Class B airspace without authorization.

My student and I inadvertently entered Class B airspace without clearance. We made a flight into Williams Gateway Airport on an IFR flight plan uneventfully and opted to depart visually for the return flight. Prior to departing Williams Gateway, I questioned my student about the Class B airspace boundaries and he stated that the floor did not begin until 6000 feet in our area. We decided that leveling off at 5000 feet would be a good plan in order to request our clearance. Having flown into the area on numerous occasions, I knew or I assumed I knew that 6000 feet was indeed the floor of the Class B airspace. However, I had taken 5 months off from flight instructing and what I did not realize was that the airspace had been redefined. What used to be the floor at 6000 feet had been lowered to 5000 feet. This change occurred approximately 2 months prior to my return. Thankfully, there were no adverse results from our transgression, other than being chastised by ATC. I believe that contributing factors to this incident were complacency on my part for assuming the airspace boundaries had not changed during my hiatus, and my failure to supervise my student more closely. As my student was checking his VFR sectional, I should have been checking it with him!

An easy mistake to make. The instructor was relying on habits developed in the past. He said, "Having flown into the area on numerous occasions, I knew or I assumed I knew that 6000 feet was indeed the floor of the Class B airspace." The instructor had failed to aim his focus on this new flight. He was the victim of having, in his own words, "Complacency on my part for assuming...."

ASRS NUMBER 426340

A Cessna 195 pilot climbs through a Class B airspace without realizing it. The departure controller advises the pilot that he was where he shouldn't be.

Departing Grand Prairie, Texas, I contacted DFW departure stating that I was off Grand Prairie en route to Conroe, Texas. I was assigned a squawk code. I reported at 3500 feet climbing to 5500 feet. The controller called back, asked if I was aware that I was in Class B airspace and to descend to 3500 feet, which I did immediately. I then realized that I had not been cleared to climb to my requested altitude of 5500 feet and had inadvertently started to climb before being cleared to do so.

A simple mistake is really another way of saying that the pilot momentarily lost his focus. In this case the air traffic controller helped the pilot switch lenses.

ASRS NUMBER 426243

The pilot of a light twin-engine airplane lands at the wrong airport while looking to follow another airplane.

I was on a training flight from 1N7 (Blairstown) to CDW (Essex County) at night. I reported 10 miles out from CDW and was advised by CDW tower to follow a Cessna on downwind. When I reported downwind as instructed by CDW tower, I could not locate the Cessna and was then advised by the controller that he would call my base. The tower put me on an extended downwind and I lost sight of airport. When I turned final, I realized I was over 10 mi out on final. When I was on a 1½ mile final, CDW tower requested my position. CDW tower stated that they had me in sight and that I was cleared to land. However, at that time I had overflown CDW and was landing at MMU (Morristown) which was 10 mi to the west. I knew that I was at MMU instead of CDW because CDW has Runway 22 as the main runway and MMU has Runway 23 as the main runway (I saw Runway 23 at the threshold). I doubt whether MMU realized that I had landed there. Corrective action: CDW tower should not have directed me onto the extended downwind behind Cessna 152. The Cessna flew the traffic pattern at about 80 knots and my Twin flew pattern at 120 knots. CDW tower should have monitored my aircraft more closely and should have never cleared me to land when clearly they were monitoring another aircraft. I should have not lost sight of the airport, but CDW tower put my aircraft on an extended downwind more than 10 mi out. I should have monitored my moving map (Argus 7000) to confirm my position. CDW and MMU are in a direct line and at night they appear similar on an approach to a night landing. When CDW tower advised they would call my base, I wrongfully assumed that they would keep me in sight and they failed to do so.

This pilot seems to want to shift much of the blame away from himself. It is not unlikely to be placed in a landing sequence behind a slower airplane. When this happens the pilot must be able to manage both the traffic separation and his own airplane. A clue was revealed by the pilot toward the end of the report that could have been a factor. The pilot says, "I should have monitored my moving map to confirm my position." Could it be that the pilot had become so dependent on the "moving map" that without it he quickly became lost? Ordinarily a "moving map" should not be required in a traffic pattern. It is easy to see that this pilot lost focus both in large and small scale.

ASRS NUMBER 386644

An air traffic controller reports the story of a pilot who either ignored instructions or was confused. Either way the pilot continued toward higher terrain despite the controller's clearance.

A small airplane was issued an IFR clearance through Louisville AFSS, along with a void time. The aircraft had filed direct GZG (Glade Springs VORTAC) and on to the SE and requested 9000 feet. I issued a route of direct ECB (Newcomb VORTAC) and climbing to 5000 feet since as filed would have put the aircraft into Indianapolis airspace without prior coordination. Direct ECB was to the northwest while direct GZG is SE. There is very rapidly rising terrain and obstructions as you proceed south and SE from K22, while to the northwest the terrain is hilly but lower. The aircraft took off and was tracking southeast when I first talked with him. He stated that he was going to ECB, and I had not yet radar identified the aircraft. As he proceeded southeast, I lost communication with the aircraft, and observed him level at 5000 feet. Another aircraft called the airplane for me and I was able to re-establish communications. By this time the aircraft was 10 nm southeast of K22, and he still said he was proceeding direct to ECB. After getting the aircraft identified, I was able to issue a clearance on course and climb him to 9000 feet. What happened? (1) The pilot figured, why go northwest when the controller will probably put me on course right away. The controller only wants to delay me...or (2) A complete loss of situational awareness by the pilot, and he was "blindly" flying along into rapidly rising terrain trying to figure out why the controller was repeatedly asking him if he is going northwest. I (we) issue clearances with very good plans in mind. While there are those times that it seems you are going out of your way, it is to keep you out of someone's way or to avoid an unplanned close encounter with the earth. If ATC does something that creates a delay or other nonsafety issue, please call us from on the ground and discuss it there.

Please do not just assume the controller is wrong (or just doesn't have the picture) and do what you want. By the way—this incident happened in deer season, not a good day to fly low over a bunch of people with guns! (grin!)

The controller attributes the problem to a pilot who thinks he knows better or "A complete loss of situational awareness by the pilot." The controller then alludes to the "big picture" from a controller's point of view when he refers to "the picture."

ASRS NUMBER 381400

A flight crew and a controller have a conflict over an assigned heading. The flight crew accuse the controller of being "at his limit."

On departure control, a controller gave us the heading off SID of 200 degrees. My captain read back 200-degrees. Four miles later the controller said "Are you on a 240-degree heading?" We said "No, 200 degrees and that's what I had read back." He then assigned us to a 270-degree heading. Then when someone on frequency stepped on someone else he said "Let me talk or we're gonna shut this whole area down." From this statement I would assume that the controller was at his limit and probably didn't pay attention to our readback.

And the captain filed his own report (ASRS Number 381401) saying "He [the controller] was clearly upset at us and instructed us to fly 270 degrees immediately."

Distraction and misguided focus can happen to controllers as fast as it can happen to pilots. The controller may very well have reached a "limit" where there was no time to switch from the big picture of the entire radar screen to the detailed requests of specific aircraft. Both pilots and controllers can get very frustrated when they know that they can do the job, but distractions keep them from completing the job. It takes patience on both sides.

ASRS NUMBER 395820

Another pilot, relying too heavily on one piece of equipment, accidentally climbed through a Class B airspace.

Remained below Class B airspace while awaiting approval to climb through at 20 DME. I started to climb from 2500 feet to 13500 feet. The controller advised that I started my climb 4 miles too soon, while I was still under Class B airspace. I stated that my DME reflected 20 miles and he replied that I was mistaken, but not to be concerned about receiving any paperwork, just watch out next time. The problem was not using my dead reckoning abilities to make sure I was outside Class B airspace as a backup to my DME.

The DME reports mileage from a VOR station. When the VOR station is not on the airport, then it cannot be used to determine distance from the airport itself. The Class B boundaries are based on radii from the airport center point. This pilot used the DME without extending his focus of attention to another source to confirm the information. The pilot admitted this when he said, "The problem was not using my dead reckoning abilities to make sure I was outside Class B airspace as a backup to my DME." The pilot lost position awareness and thought he was cleared to climb when in fact he was not.

ASRS NUMBER 389840

A student and flight instructor making repeated touch-and-go landings at a busy airport fail to get landing clearance on the fourth approach. They land anyway between departing and arriving traffic.

Late in what would be a 1 hour training flight in the pattern at San Jose International Airport, my student and I did a touch and go on Runway 11 without clearance for the option. Winds during the flight at the surface were variable from 130–210 degrees at 9 knots variable 16 knots gusting to 21 knots. Winds aloft at the traffic pattern altitude of 1,000 feet MSL, were much stronger and at a direct crosswind. Needless to say, the patterns and landings were difficult for my student and there was light turbulence on the downwind. I continued the flight, however, because my student was capable of handling the aircraft in these conditions and it was excellent experience for him. After approximately 4 touch and goes on Runway 11, my student told me on the upwind that he thought we had not been cleared to land. He was right, we had not been cleared to land, or for the option, for the previous touch and go. I had completely forgotten and tower said nothing, so I told my student to say nothing over the radio and continue as before. Tower cleared us for the option approaching base on the next landing and the flight continued without further incident. While not too dangerous by itself, our uncleared landing was dangerous considering how busy the airport was. Tower departed 1 aircraft off of Runway 11 while we were on final, and another was landing behind us, there was the usual steady stream of airline departures and arrivals on Runway 12R, and several aircraft were stacked up on Runway 12L for departure. The tower simply forgot about our Cessna in the pattern, and we forgot to get cleared to land as we were busy with the difficult patterns and watching for traffic. Not exactly my finest moment as the all-powerful, student-mistake-catching CFI.

Everyone lost focus during this scenario. The controller and flight instructor just forgot, and the student did not speak up until after the fact.

Lost and Misplaced Pilots

ASRS NUMBER 425638B

A student pilot entered the Windsor Locks Class C airspace after becoming disoriented on a solo cross-country flight. The student realized his mistake but did not contact the Windsor Locks approach control.

I was a student pilot on a solo cross country, flying from Scotia, New York, to Providence, Rhode Island, and became disoriented and drifted south into the Windsor Locks Class C airspace. I reoriented myself and continued on to Providence without contacting Windsor Locks approach. This was a mistake. I was at 3500 feet MSL and below the 4200 feet ceiling of this Class C airspace. Upon landing at Providence I was instructed by the tower to call Windsor Locks (BDL) approach, which I did. The supervisor at BDL explained what I had done and asked me how it happened, I explained. He then explained how I should have contacted approach when I was disoriented and that they were there to help. I apologized and thanked him for making me contact him. This was a major mistake on my part and thankfully no incident occurred. I know the seriousness of this situation and have learned from this experience and will be more aware of airspace violations in the future. I have also spoken with my CFI and received instruction on what to do when lost and how to prevent future violations.

ASRS NUMBER 425295

A traffic conflict takes place in the traffic pattern at Athens, Georgia. A student pilot and his instructor report an inaccurate position, which made the controller think there was enough time to issue a takeoff clearance to another airplane. The instructor and student, being closer to the airport and coming in from a different angle than requested, have a near collision with the departing aircraft.

We took off for Athens, Georgia. Five miles out, I called Athens (AHN) for landing. Instruction from AHN: Call 3 miles west of Athens. My student said "3 miles" to AHN tower. The real position was more like 2 miles out. The tower instructed "Enter left downwind Runway 27." My student proceeded with turn into left crosswind! Apparently, tower had authorized another airplane on takeoff for an early right turnout, which resulted in a near miss. I took airplane from student pilot and initiated evasive action to left (seemed direction in which greatest separation in shortest time could be obtained). While this was happening, tower spotted us (I believe) for the first time and instructed us to a right downwind. I gave the airplane control back to the student after asking if he understood. Student said yes, and we proceeded. The student turned onto final for Runway 20, while yet another airplane was on final to Runway 27. He notices this and aborts landing. I am flying the plane in attempt to re-enter downwind Runway 27 but AHN tower reacts right away with instruction to land on Runway 20. We are landing on Runway 20 as instructed. Tower asking for meeting. Visited tower manager. No further action after talk.

ASRS NUMBER 427760

The pilot of a small airplane loses his position awareness and crosses through a Class D airspace without knowing it.

While descending into Chino, California, on a VFR flight from Visalia, California, I apparently violated Brackett's airspace. I called Chino approximately 10 miles out after listening to ATIS, received landing instructions, landed and then was told to call Brackett ATC. I was informed by the ATC controller I had navigated through their airspace. I was surprised to hear that, as my plan was to fly under the outer ring of the ONT Class C airspace through the corridor between Brackett and Ontario. I told the ATC controller this was my first trip into this area, I had a current map in front of me, with an Argus 7000 moving map, an additional handheld GPS and another pilot as an observer with me. We were doing everything we could to do what we thought was correct, and that I was sorry as I never saw them. I heard the ATC controller that I was talking with make a comment to someone else in the tower, explaining the situation as it happened and that I was new to this area. The person in the background said "Just get his information and write him up" in a rather short fashion. I provided all the requested information to the ATC controller. I was also told that traffic had to be redirected. This, however, I don't understand, as both myself and my passenger (pilot) were scanning continually and never saw another aircraft anywhere close. As I was approaching Chino, I thought I was looking out my left side at ONT (a double parallel runway) and I needed to stay west of a north/south road which is west of the runway. Obviously, now, it was not the correct road and not the correct parallel runway. In the haze and ground clutter, and since it was my first time into Chino, my ground reference must have been incorrect. I have flown into 2 different countries, into the Arctic Circle, and I cross the United States coast-to-coast several times a year and I have never violated anyone's airspace before. I was surprised to hear what had happened.

ASRS NUMBER 427736

Have you ever just had one of those days? This student pilot ran into trouble twice on the same flight. First he lost his position awareness and flew into airspace of a neighboring airport without authorization, and then when he was able to get back to the proper airport, he actually flew under another airplane on short final.

I am a student pilot and was on a long solo cross country. I was on approach to Ann Arbor, Michigan (ARB), I contacted the tower and asked for permission to land. I was about 12 miles to the southwest, or so I thought. They responded by telling me to report in when I was on a 4 mile final for Runway 6. There was very heavy radio traffic and airplane traffic at ARB. It was very hard for me to get a break in which to talk with the tower. I began looking for the airport and Runway 6. I saw in the distance to my northeast what I believed to be ARB Runway 6. When I was close enough to announce to ARB tower I was 4 miles out,

I was unable to get my radio transmission over the other radio traffic. At that time I realized the airport was not ARB, but Ypsilanti (YIP) and that I had violated their airspace unintentionally. I quickly turned to the west and headed for ARB (which is only a few miles away). When I was out of YIP airspace, ARB called and told me I had gone into YIP airspace, and asked me to continue back to ARB on a right downwind for Runway 6. I looked at my sectional and saw that YIP has a Runway 5, which is why I must have mistaken it for ARB Runway 6 at a distance.

The landing incursion happened while I was going back to Ann Arbor (after entering YIP airspace by mistake). ARB tower told me to turn after the Cherokee that was on a left downwind for Runway 6 (I was on a right downwind). I saw the Cherokee on the other side of the airport and reported to the tower that I had traffic in sight. The radio traffic was nonstop between the tower and aircraft, and I don't think they received my transmission because they came back and asked me if I saw my traffic to follow turning base. I responded that I did. At that point I turned base after giving the Cherokee some time on final to make sure we had good spacing for landing. I estimate that I was 3–3½ mi out and the Cherokee was 1 mile ahead of me. I was maintaining pattern altitude (1500–1600 feet MSL) and was watching the Cherokee ahead of me. The Cherokee started getting really low (about 200–300 feet above the tree tops) and I didn't know what he was doing (there was no radio communications from the Cherokee to the tower, but the radio traffic might have prevented that). Then the Cherokee started getting higher and when he crossed my level of sight, I could tell I was gaining on him. He then went up to about 2000 feet or 2200 feet and we were about less than 1 mile from Runway 6. The Cherokee had clearance from tower to land, but it did not appear he was landing. I was trying to radio the tower to ask for spacing maneuvers, but the radio traffic kept me from getting through, and by that time I was getting very close to the Cherokee, so I lowered my altitude and passed underneath him, at which point the tower saw what I had done and cleared me to land ahead of the Cherokee. I don't know what the Cherokee was doing, because I was in landing configuration only doing about 80 knots. I didn't know why I overtook him. I know that next time I will not wait for the tower to respond to my spacing, I will do maneuvers as needed and inform the tower afterwards when there is so much radio traffic.

ASRS NUMBER 427189

This Cessna 172 pilot became disoriented and confused after flying into an active Military Operations Area by accident. VFR flight into an MOA does not require special authorization, but the pilot, working with center controllers, became confused by a series of radar vectors.

I departed Lake Havasu City, Arizona (HII), flew up the Colorado River to Needles VOR. I opened my flight plan with Riverside radio 122.2. Flying a Cessna 172A. Then I contacted Los Angeles center for a transponder code. My destination was Fox Airport (WJF) in Lancaster, California. I was flying a heading of 255 degrees to the west and found myself in a military operations area (MOA), not knowing it was active. The controller gave me many different headings to the point that I told her I was getting confused, trying to fly the airplane and read the charts to get out of area. After a few headings she told me to contact the center when I got to Fox Airport. I closed my flight plan after arriving at the Fox Airport. I called LA Center and talked to the man about what happened in that area and he said I would receive a letter or be contacted about the airspace I entered. I feel like I should have some warning when I first contacted LA Center out of Needles VOR.

ASRS NUMBER 426421

Another pilot loses his position awareness, and this leads to problems. Before he discovers where he is by seeing a prominent landmark, he inadvertently flies through the Seattle Class B airspace twice.

On a cross country flight from Arlington, WA, to Tacoma, WA, I became disoriented, and while trying to find my position, I may have entered Class B airspace. While at Lake Tapps, I turned towards my destination on the other side of McChord. I thought I must have drifted south because I didn't see McChord until they told me I was 8 miles northeast, which may have put me in Seattle's Class B airspace—but I was so frustrated I didn't realize even then where I was, and by then I was heading west over Vashon Island, then climbed to 4000 feet MSL to see better. At that time I identified Tacoma Narrows Bridge and knew where I was. I made a poor judgment of my position and didn't use all the resources available. My lack of confidence in my ability caused me to freeze and not think straight. I'm going to receive additional instruction and practice to make sure I know where I'm at, at all times.

Pilots Unaware of Dangers That Surround Them

ASRS NUMBER 426044

The classic VFR and IFR collision setup. This report is from the pilot of a Piper on a straight-in ILS approach. Meanwhile, the pilot of a Cessna, flying in the uncontrolled airport traffic pattern, makes a turn from a right base to final. In this case the Piper (low wing) ended up just 200 feet below the Cessna (high wing) while both were on final to the same runway. Both pilots lost their awareness of the traffic pattern situation at a critical time. Had it been the other way around, the low wing above and the high wing below, this might have been an accident report instead of a ASRS report.

The problem arose when the conflicting aircraft entered on a right base while I was on final from an ILS practice approach during VMC. There was another aircraft behind me on the ILS. The Cessna 210 on base called in sight and thought it was me, so he really never had me in sight. I could not make any calls when 3 and 4 mi out because the Cessna 210 was talking to UNICOM about something other than flying at hand, so I couldn't talk with him. I noticed him right on top of me about 200 feet and immediately turned right and dove away. I noticed him right when he stopped talking to UNICOM and announced short final Runway 2. Two aircraft on final were contributing factors while the Cessna 210 was blocking the frequency talking of other matters than the ones at hand.

ASRS NUMBER 425237

The pilot of a Cessna 172 ignored the forecast for icing and took off into the clouds. He believed that he could "dodge" the ice by flying between cloud layers. As the flight progressed, the pilot got deeper and deeper into trouble before the danger really dawned on him.

During an IFR cross country from LUK to MCN, we encountered light icing about 20 nm north-northwest from the LOZ VOR. The wings looked just slightly frosty, almost as though someone had been breathing on the wings. There even appeared to be a slight "shadowing" of the wings. We exercised good CRM and immediately decided to divert to the nearest airport, which thankfully had a VOR approach. However, while we descended from the MEA for the instrument approach, we picked up moderate mixed icing. The temperature at the MEA (5000 feet) was 30°F. I assumed that a descent of 3000 feet for the approach would prevent additional ice from forming. Was I in for an unpleasant surprise! We quickly picked up between ½–¾ of an inch on the entire airplane, and the temperature gauge never moved from 30°F during the entire approach. I am writing this form to admit some mistakes I made and reveal some things I learned. I was wrong to think that I could fly through an area forecasted to have light to moderate rime icing in the clouds.

I assumed that because layers were reported that I would easily find an altitude that would keep me out of the clouds. I put myself in a situation where I felt I would have to declare an emergency before it was too late. I could have never imagined how quickly the airplane went from "feeling like" a Cessna 172 to "feeling like" a heavy Boeing 747. We were a very heavy C172. We took off at maximum gross wt, 3 passengers and full fuel in a Cessna 172 with extended fuel tanks. We had a GPS on board that really allowed me to see the effect the ice had on our airplane. As we picked up more and more ice, I noticed that the ground speed indicated on the GPS continued to decrease. Another thing I learned is that it's really easy to spot ice accumulation on the black tires. I will never fly in known or forecasted icing conditions ever again.

ASRS NUMBER 425015

A student pilot, flying solo in preparation for an upcoming flight test, hits a cable. The student was making a "practice forced landing" and obviously got too close.

I was flying solo over the area of Carbondale, in a Cessna 172. The intent of the solo flight was to satisfy the last 2 hours of my prerequisite and to practice for my FAA flight test. I was airborne from Glenwood Airport. Climbing to an altitude of approximately 9500 feet MSL over the Carbondale area, I practiced stalls, figure eights, S-turns, and emergency landing procedures. After a couple of simulated emergency landings in open fields and an unused runway I recalled that I had been told that the FAA inspector preferred simulated emergency landings on roads. I then proceed to simulate an emergency over a local hwy. My intention was to see how a plane could land on a road if there was traffic. I selected the side of the highway with no utilities and flew parallel to the highway watching traffic speed against the aircraft optimum glide speed. When sure that I could land safely I noticed a car pull into the traffic lane where I envisioned my plane landing so I hesitated my ascent as I contemplated this change of events. As I pulled the nose up to climb away I saw a flash out of my left eye and felt the plane shudder momentarily and then continue the ascent. The light utility line I connected with was running across the road and the position was on a high bank (approximately 500 feet bank) to my right. I was not aware of this because I was concentrating on the car. I returned to Glenwood Airport and landed safely. There did not seem to be any damage to the aircraft except possibly a mark on the passenger side strut cover, which my instructor was not sure if it had been there before the incident. My instructor immediately debriefed me on the dangers of low flight. I am very aware of the potential for disaster in this incident and will now simulate my emergencies at a safe altitude. I am also more aware of the operating heights around obstructions.

A side note on this story. Usually a field is better as a forced landing site as opposed to a highway with cars on it. You never know what the driver of a car will do when they are surprised by an airplane, and highways very often have cables and wires crossing over the top. I always advise against student pilot solo practice of emergency procedures, and the 500 feet above people, cars, objects, etc., should always be followed.

All of these examples were of pilots who lost some portion of their awareness, and due to this lapse a chain of events began that lead to trouble. If any of these pilots had rediscovered an awareness of the situations they were in, the problems could have been avoided or at least reduced. It seems that once awareness is lost, it is very hard to regain. It would be better to maintain your awareness in the first place and by doing so preempt the problems. The next chapter is a guide to maintaining your awareness.

3

Maintaining Situational Awareness

Situational awareness is a moving target. If you have it one moment, you may lose it in the next moment. You may have it one moment and in the next moment you may lose it without knowing that you lost it. Maintaining situational awareness is a constant, never-ending effort. The aircraft is in constant motion, therefore the situation that is presented to the pilot is also ever changing. To have any hope of keeping up, the pilot must learn to anticipate changes and formulate contingency plans.

Even when flying on a beautiful VFR day, the pilot must become the receiver and processor of a large amount of incoming data. In flight, all the pilot's senses are feeding information to the brain. The eyes scan the instruments and the horizon, the ears listen to the rhythm of the engine and ATC callouts, and even the "seat of the pants" detects every air current and eddy. The pilot must become "tuned in." But situational awareness implies more. In addition to simply concentrating on the surroundings, the pilots must also be suspicious, savvy, and calculating. The pilot must bring in all this outside data and then make sense of it. The pilot must look for clues of an impending problem. Aware pilots anticipate trouble and conduct preemptive strikes on problems. Unaware pilots only learn of a problem's existence as it is unfolding and when it is too late for remedial action.

Aware pilots are proactive decision makers. In fact, the state of awareness and assertive decision making depend on each other. They are like binary stars in orbit around each other. Decisions that pilots make tend to reveal additional information. This new information is then used to enhance awareness. Then with greater awareness, better decisions are possible. It is a positive cycle.

Pilot decision making is not static. When a decision is made and the outcome of the decision has no effect on any future decisions, it is called a static decision. But when we fly we know that the outcome of just about every decision will change the available options and influence the next decision. It is like coming to a fork in the road. If you select the right fork, you will, by your decision, eliminate any of the challenges that you would have faced if you had selected the left fork. In an airplane, once a course of action has been selected you probably will be unable to stop, back up to the fork in the road again, and select the other fork. In an airplane the decisions we make are dynamic. Decisions line up one after another in chains. There is not just one fork in the road, but a succession of forks, each one changing the options and producing different outcomes.

Knowing the best fork to take depends on the pilot's ability to make a good situational assessment. This assessment depends on the pilot's ability to project or forecast the result of the decision. The ability to project and forecast is more often referred to as "staying ahead of the airplane" by flight instructors.

I have conducted research with groups of instrument-rated, general-aviation pilots. In one project, pilots flew an IFR flight in a flight training device (simulator). At one point in the flight, the pilots flew a localizer-only instrument approach. The pilots did not know it at first, but the cloud bases were lower than the minimum descent altitude (MDA) for the approach. Most of the pilots correctly flew down to the MDA and discovered that they were still in the clouds. No runway ever was sighted, so when they arrived at the missed approach point, they properly executed a missed approach. When these pilots were then asked, "What are your intentions?" most asked for vectors to another approach at another airport. They reasoned that the ceiling was too low at the first airport and so another attempt there was a waste of time. But several other pilots did not fly the approach correctly. Some managed the approach so poorly that they never got all the way down to the MDA before reaching the missed approach point. Because they did not get down to MDA, they did not know that the clouds were lower than the MDA. When these pilots executed a missed approach and were asked, "What are your intentions?" most of these pilots asked for vectors back around to shoot the same approach a second time. They reasoned that if they had made it all the way down to MDA, then they would have seen the runway and could have landed—when in reality, the clouds were too low and a second attempt at the same approach was a waste of time. So the poorly flown approach robbed them of valuable information, made them less aware of the true situation, and adversely influenced their next decision.

There was one more factor that the pilots were also dealing with during this localizer-only approach. During that portion of the simulator session, the alternator had failed and the airplane was using battery power only. The pilots who knew a second attempt was going to be a waste of time quickly diverted and landed safely using a nearby ILS approach. But the pilots who were unaware of the total situation and asked for a second attempt back to the same airport using the same localizer-only approach soon used up all their battery power before making a final decision to divert. A lack of situational awareness was fatal in the scenario. Fortunately, this experiment was done in the safety of the simulator and everybody walked away unhurt. But the interdependence between pilot flying skills, situational awareness, and pilot decision making was never more clear.

An accident in 1996 took place that was very similar to my research scenario. A private pilot, flying in instrument conditions, was killed while attempting to land at San Diego's Brown Field.

NTSB REPORT LAX97FA049 The pilot got two weather briefings and was advised of deteriorating IFR weather in the destination area. When he filed an IFR flight plan, he told the briefer that he did not have his charts; the briefer looked up the airway designations and fixes for the pilot. Near the destination approach control told the pilot the airport was below approach minimums and that three other aircraft had made missed approaches without seeing the ground. The controller then suggested nearby airports that were above approach minimums as alternatives. The pilot said that his car was parked at the airport and he wanted to make the approach. Radar data disclosed that the aircraft flew the approach segments at least 1,000 feet higher than the charted altitudes and at speeds between 155 to 180 knots. The aircraft overflew the missed approach point and the airport, then crossed the adjacent US/Mexico border before ATC could instruct the pilot to make an immediate missed approach. The pilot responded on the radio "I guess I don't know where I am." Radar data then showed the aircraft climbing and descending rapidly as it reversed course, then descended to 300 feet above the ground as it neared the western airport boundary. The pilot transmitted that he thought he had the airport in sight. Four seconds later, the aircraft impacted the departure end of the runway. Ground witnesses observed the aircraft in cloud bases, and they noted that it narrowly missed a building; it then turned sharply toward the runway before descending steeply to ground impact. The pilot's logbook did not show that he had met instrument currency requirements of FAR Part 91.

The NTSB's official finding of the probable cause was: "The pilot's lack of situational awareness, his failure to fly the approach as charted, and his failure to maintain aircraft control, while attempting an abrupt turn toward the airport, which led to an inadvertent stall/spin. The pilot's lack of recent experience was a related factor."

There were some differences between this accident report and the simulator scenario, but they did have this in common: When pilots flew the approach poorly, poor decisions followed. This was, in part, because the pilot did not acquire all available information and this limited their awareness.

The pilot in this accident report was given information about cloud bases and alternate airports but chose not to consider these factors. The pilot repeatedly made poor situational assessments.

Making a poor assessment separates a pilot from situational awareness, and this can lead to either an overreaction or an underreaction to the situation.

NTSB report LAX89LA274 tells the story of a pilot who overreacted to a situation because of his failure to properly understand the situation.

The pilot of the accident aircraft reported smelling and see-ing smoke in the cockpit area, coming from under the pas-senger side dashboard. The smoke lasted for about 20 seconds. The pilot decided to turn off the throttle, mixture, and fuel and execute a forced landing. He also reported hearing the stall warning indicator 2 to 3 times during his emergency descent. The aircraft landed hard on a dirt road.

The pilot was seriously injured and the airplane was destroyed by the impact. The NTSB statement of proba-ble cause: "The pilot's failure to maintain airspeed dur-ing an emergency descent and to properly flare the aircraft during his landing. Contributing to the accident was the pilot's anxiousness in overreacting to a situation and prematurely attempting a power-off landing."

The pilot misdiagnosed the problem. In reality the smoke was from smoldering electrical wires under and behind the instrument panel. The presence of smoke was short—20 seconds. The engine and its operation have nothing to do with the electrical system. This Piper air-plane, like most other airplanes, had a magneto system for engine ignition. The engine was not the problem, but the pilot shut down the engine anyway. A more appro-priate situational response might have been to turn off the master switch. This would have prevented the flow of electrical current throughout most of the airplane and probably would have eliminated the problem. The engine would have continued to run with the master switch off. This accident took place during daylight hours so the pilot could have flown to the nearest airport safely. The fact that the pilot did not fly to an airport, but instead elected to make a power-off, emergency, off-airport land-ing was the result of an incorrect assessment of the situa-tion. The pilot must have believed that his life was immediately in danger from a fuel fire. He became unaware of the true situation and as a result overreacted.

The pilot involved in NTSB report LAX98GA029 did not act soon enough and did not maintain vigilance resulting in a fatal accident of a Cessna 182.

The aircraft impacted mountainous terrain while on a Civil Air Patrol search mission. The surviving passenger stated that the pilot-in-command allowed the front seat observer, who was a pilot, to fly the aircraft while the pilot-in-command scanned the terrain to the left of the flight path. The observer/pilot allowed the aircraft to get too low and too slow for the altitude and terrain conditions, and the pilot-in-command was unable to take over in time to prevent impact with a pine tree that caused the aircraft to nose down into the ground.

The NTSB concluded that accident as caused by "the pilot-passenger's failure to maintain sufficient altitude and airspeed to avoid a collision with the terrain, and the pilot-in-command's diverted attention and failure to recognize a deteriorating situation in time to effect a safe recovery."

So maintaining situational awareness is essential to proper situational assessment. And proper situational assessment is required to take the proper remedial action. Pilots who do not maintain their awareness cannot assess and therefore can underreact or overreact.

There are also situations where the pilot does not act at all and/or misinterprets information that could have maintained awareness. This can start a runaway chain of events.

FIGURES 3-1 and 3-2 show the sad ending to a private pilot's VFR flight. Almost miraculously the pilot in this airplane climbed out with only a scratch on the back of his hand. The pilot was attempting to make an off-airport precautionary landing. In FIG. 3-1 the pilot's approach was from the left of the frame. On final approach the airplane came into contact with the power

3-1

3-2

lines on the left. If you look closely you can see where the power lines have been repaired by splicing additional wire into the line. The airplane first contacted the power line that is farthest to the left in FIG. 3-1. The airplane hit the wire at a slight angle and for a moment slid down the wire. But the airplane's momentum carried it and the first wire through to contact the second and third wires. When the three wires came together, there was an enormous orange flash, a loud buzzing sound, and that unmistakable smell of burned electric insulation. When the wires came into contact, the electric current in the wires shorted out and power went out for an entire eastern Kentucky county. The airplane's forward motion was all but stopped as the power lines burned and gave way. The airplane then fell, nose first, about 50 feet from the height of the power line to the field below. FIGURE 3-1 shows the height and angle of the final drop. The nose hit pointing straight down and the remainder of the airplane's forward motion carried it over on its back. FIGURE 3-2 was taken the morning after the accident as the crew repaired the power line, and shows the airplane's final position.

When this airplane crashed, its engine was operating properly, it had more than two hours of fuel remaining, its electrical system, radios, and avionics were later tested and found to be in proper working order, and there was an airport less than 10 miles away. In other words, this pilot flew a perfectly good airplane into the power lines when there were at least half a dozen better options. What could have led a pilot into a situation where he believed that taking the chance on an off-airport landing was the best course of action? This case has become a classic example of an accident caused directly by the loss of situational awareness.

The NTSB report gives only the surface information. NTSB identification number NYC99LA067 reports:

On February 26, 1999, about 1945 eastern time, a Cessna 152 was destroyed when it contacted wires while attempting an off-airport precautionary landing, near Sublett, Kentucky. The certificated private pilot received minor injuries. Visual meteorological conditions prevailed for the cross-country flight destined for Raleigh County Memorial Airport in Beckley, West Virginia. A visual flight rules flight plan had been filed and activated for the flight conducted under 14 CFR Part 91.

The roots of this accident started to take hold several weeks before the actual date of the accident. The private pilot was trying to accumulate VFR cross-country flight time so that he would eventually qualify for the instrument rating. He had originally planned the flight to West Virginia more than a month before the accident flight. He wanted to combine the cross-country flight requirement with a visit with relatives. Between the winter weather and busy work schedules the flight was canceled several times. On the day of the accident, the pilot was working as a lineman at the departure airport. He saw that the weather was good and an airplane was available, so he planned to switch work shifts with another employee and take the flight that day. He called his relatives and told them that he would be making the flight that day and arranged for them to meet him at the Beckley airport.

He tried but failed to get another employee to come in and cover his shift. He asked to get off work early, but it was a busy day and with nobody else to do the work, he was told he had to finish his schedule if he could not get anyone else. For the rest of the afternoon he simultaneously worked the flight line and planned his flight. He made one more call to the relatives in

Beckley to tell them that his departure had been delayed, but that he would soon be on his way.

At some point during the afternoon he spoke with another pilot and invited the pilot to ride along on his cross country to Beckley. The passenger/pilot agreed and they made plans for their departure. At the last minute, however, the passenger/pilot decided not to ride along. Later, when the passenger/pilot was interviewed he said that, "It just didn't feel right." He had actually sat in the airplane while the pilot started the preflight inspection. The pilot had been interrupted several times while he attempted to preflight so that he could fuel other airplanes. The pilot seemed rushed, preoccupied, and was interested in getting away fast. The passenger/pilot saw no preflight navigation plans, and no sectional chart. Eventually this uneasy feeling led the passenger/pilot to get out of the airplane and decide not to go along.

Finally the pilot took off, alone, around 4:00 p.m. On the climbout he contacted Flight Service and activated a VFR flight plan and later called the Air Route Traffic Control Center for flight following. During the day he had used the Direct User Access Terminal (DUAT) to obtain a weather briefing. While online he used the DUAT flight planning option to calculate a course over three VOR stations and on to Beckley. He printed out from DUAT the VOR frequencies, a time en route, and a fuel burn estimate. He was using that printout as his sole navigation reference. He did not have a sectional chart so he did not select any checkpoints to use as references along the way.

He climbed to 5500 feet and passed over the first of the three successive VOR stations. Between the first and second VOR the center gave him a frequency change. At some point between the VORs, his heading began to wander off course and he was unable to rejoin the proper VOR radial. He made several attempts to intercept, but he misread the VOR indicator and at one point was 90 degrees off his heading. He had relied completely on VOR navigation and now he was having trouble understanding the VOR indications. With no chart to use as a backup, he quickly became disoriented.

While he wrestled with the VOR and tried to understand where he was, his altitude began to inadvertently decrease. Later the pilot said that when he made the decision to make the off-airport landing that he was at 3500 feet. So while he was attempting to regain his position, he allowed the airplane to descend approximately 2000. He was over the eastern part of Kentucky and over some very rugged Appalachian terrain. Indianapolis Center had been given the hand-off for the flight, but they were unable to receive any transponder signal. Later it was determined that due to the mountainous terrain an altitude below 3800 feet MSL would have made transponder and VOR reception difficult or impossible. When Indianapolis told the pilot that no radar target was observed, they asked him to "recycle" his transponder. The pilot turned the transponder off and back on several times, but the radar target was never seen by Indianapolis.

Now at 3500 feet the pilot's VOR receiver started to show an "OFF" flag and the course deviation indicator (CDI) became erratic. The pilot reported later that the "reply" light on the transponder was "dim." Most transponders, including the one in this airplane, have a feature that will raise and lower the brightness of the reply light. This allows the pilot to turn down the brightness so that at night the flashing light is not distracting. The pilot was not aware of this feature and thought that the dim light meant that his electrical power was low.

His VORs were not working, and Indianapolis could not receive his transponder. He concluded from this that his electrical system was failing or had already failed.

The pilot did not hear Indianapolis for a time and believing that his radios were failing, he began switching radio frequencies. He made calls on random frequencies. He tried using the airplane's hand mike instead of his own headset. He recycled the transponder a few more times, and while doing this he inadvertently left the transponder select knob in the "off" position.

Later the pilot said that he looked at the airplane's ammeter to determine the condition of the electrical system. The ammeter normally displays whether or not the battery is charging or discharging. The pilot said that the ammeter read "zero." This would mean that the battery was not accepting or dispensing a charge. In normal flight and under normal conditions the ammeter will read zero or slightly past zero on the positive side as the alternator keeps a "trickle charge" on the battery. But the pilot misinterpreted the ammeter reading. At that anxious moment he thought the zero reading meant zero electric power and concluded he had no electricity.

At this point in the flight what was actually taking place and what the pilot thought was taking place were on divergent courses. The pilot believed something that was untrue. Pilots have a term for how the mind can play tricks on you when you are in a stressful situation. It is called "automatic rough." When you fly out over open water, across a desert, or over mountains you can get anxious. You automatically hear the rattles and vibrations that were probably there all along but now you focus on them. Soon you have yourself believing that something is wrong. The stressful situation can convince your mind that you have trouble when you don't. This situation can launch a separation between what is real and what is not real. Another definition of situational awareness could be the state of believing what is true and disregarding what is not true. Usually we think of a lack of situational awareness occurring when something bad is happening that the pilot is not aware of yet. But a loss of situational awareness can also take place when the pilot believes something bad is happening when it actually is not.

The pilot was now flying the airplane under a false set of assumptions. Every decision that he made downstream from this point was flawed because he had lost his awareness. Reality went one way and his mind went another.

From this point on he also had an underlying sense of panic. He was completely lost. He was over rugged terrain. It would soon be dark and he thought he had no electrical system. Every thought he was having at this point was filtered through panic and this made things go from bad to worse.

In a calm, controlled, classroom-type situation he might have remembered that the brightness of the transponder reply light is not an indication of its power output. If he could have just "parked" the airplane for just a minute to think, he might have remembered how to interpret the ammeter. Without this underlying panic he might have remembered to climb, because the higher you go the better radio reception you get. But an anxious situation can tie your brain in knots. How many times have you said to yourself, "What was I thinking?" You have been in situations before when a combination of time-urgency and stress has made you act differently than if you had the luxury of time and deliberation.

As the sun got lower in the sky, and he got even more lost, he started to believe that he was running out of fuel. He later reported that he observed the right fuel tank indicating near empty and the left tank indicating only one-half tank. The airplane had been filled with fuel to its capacity before departure. He had been in the air less than two hours and his DUAT fuel calculations indicated that he still had more than two hours of fuel remaining. But the gauges were swaying back and forth and he started to believe that despite the calculations, fuel exhaustion was imminent. After the crash and despite the airplane coming to rest inverted, an FAA inspector drained 5 gallons of clean fuel from the left tank and 6 gallons from the right tank. The fuel calculations had been essentially correct. He had enough fuel to have flown back to where he first took off.

But once the chain of events started to unfold there was no turning it around. The pilot prepared poorly and set himself up to get lost when he did not take a chart and relied on VORs alone. He was rushed into the flight with family pressures and thoughts of past delays driving his decision to go. He got out over rough terrain and it was as if he was over the middle of the ocean: There were no discernible landmarks and no place to land. Panic started slowly but crept into the back of his mind. Soon the panic was affecting his ability to reason and think.

Look at how deeply the loss of awareness had developed. The pilot thought his radios and transponder were broken when a short climb would have made them work again. He thought his electrical system was not working but he was relying on fuel gauges, which operate only with electricity. He thought he was running out of fuel when his watch, his scant fuel calculations, and common sense said otherwise. FIGURES 3-1 and 3-2 both show that the airplane crashed with the landing flaps in the down position. The flaps on this airplane are moved by an electric motor. The fact that the flaps came down should have told the pilot that he had electric power available.

His final decision was driven by the fact that he was situationally unaware. He thought that he would soon be flying in the dark with no lights, over the mountains, with the engine running on fumes. He was more than 50 miles off his original course but he was less than 10 miles from an airport. Flying too low to triangulate any VOR stations and with no chart to match up landmarks, his only hope would have been to accidentally fly over a runway. That didn't happen. Instead, he spotted an open area running along the base of a valley. It was the first flat area he had seen in over an hour. He reversed course to get back to the valley and flew around one time before entering what was essentially a downwind leg to the valley. The sun was low now and in the shade of the valley he never saw the power line, flying over it once on downwind, parallel to it on left base, and into it on final.

The pilot had failed to maintain his awareness of the situation and almost killed himself because of this failure. He did walk out of the field and across to a house. The people at the house wondered where this guy had come from but mainly they wondered why they had no electric power. They soon discovered that the two were connected.

A similar, but more tragic, accident happened near Muleshoe, Texas. NTSB report number FTW96FA143 tells the story:

The student pilot departed on his initial solo cross country flight at 1500. He became lost on the first leg, and after receiving navigation assistance from ATC, made a full stop landing at his planned interim airport about 2 hours and 16 minutes after departure. A ramp attendant said that the student seemed "very nervous." The ramp attendant asked the student if he wanted to refuel, but the student refused. The student departed the interim airport at 1750. About 50 minutes later, he became lost again. Pilots of other aircraft and ATC attempted to help the student via radio, but were not successful. When asked if his VOR was functioning, the student replied, "it's out of order, [and it] says OFF and I can't get a TO or FROM." After nightfall, the airplane hit a powerline and crashed on the shoulder of a highway about 60 miles north of the intended destination. No fuel spillage was detected at the site, and only about 1 gallon of fuel was found in the fuel tanks. The total flight time without refueling has about 4 hours and 21 minutes; the last 30 minutes at night. According to performance charts, fuel consumption would have been about 5.6 gallons per hour. Fuel capacity was 26 gallons of which 22.5 were usable. The pilot had a total of 27.6 hours of day VFR flight time, and 1.6 hours of solo time. No preimpact mechanical defects were found. No record was found that the pilot's flight instructor or flight school had contacted FAA authorities to inquire about the overdue aircraft.

The student was killed in the accident. The NTSB's probable cause statement aimed the blame at the student's inability to maintain situational awareness but also at his flight instructor's inadequate awareness training. The NTSB concludes that what probably caused the accident was the "Improper planning/decision by the student pilot, and his failure to take adequate remedial action after becoming lost and encountering a low fuel situation, which subsequently resulted in fuel exhaustion, loss of engine power, and a forced landing at night. Factors relating to the accident were: The student pilot became lost/disoriented, inadequate supervision by the flight instructor (CFI), darkness, and the inability of the student to see the powerlines during an emergency landing at night."

The loss of situational awareness can creep into any cockpit. The last two examples were of inexperienced pilots in light, general-aviation airplanes, but experienced flight crews in large jet aircraft are not immune.

One of the most intriguing cases that ever exposed the loss of situational awareness took place when an airliner accidentally landed not only at the wrong airport, but in the wrong country. Aviation Safety Reporting System report number 315108 was written by the crew's First Officer and tells the story:

I was the pilot-not-flying en route to Frankfurt, Germany. A descent from flight level 350 was issued to us from London control and this was initiated near the Biggen VOR. We received further clearances to descend until reaching FL180, when we were handed off to Brussels control. Brussels requested that we expedite our descent to FL160. At this time they began calling us by a different call sign, but we thought this was for their clarification. We commented among ourselves that we were getting down very early for a descent into Frankfurt. After receiving clearances through 12,000 and then 10,000 feet we were cleared to the Bruno VOR which we could not find on our map. We asked ATC for assistance and vectors were issued. At about this time, the Second Officer located the Bruno VOR on the chart, and we obtained a 110.6 frequency. I began looking at the Frankfurt approach charts. Coming out of 8,000 feet, the Second Officer became busy with attempts to get ATIS and contact the company. She had to verify that the Number 2 communications radios were working and repeatedly attempted to use a faulty company frequency to contact dispatch. The Brussels controllers were extremely difficult to understand. We had to ask them for repeat clearances. We were advised to expect a runway 25L approach, and so the Captain briefed a runway 25L approach to Frankfurt. The ILS, ADF, and DME frequencies for Frankfurt were set in the appropriate radios. We changed over to another approach controller and cleared down to 8,000 feet and then 4,000 feet. After passing Bruno VOR, a vector for runway 25L was given. Thinking we were getting close to Frankfurt and that we were talking to Frankfurt approach, we continued to attempt positive identification for the runway 25L ILS approach. We also continued trying to contact ATIS and the company. The runway 25L ILS at Frankfurt with a frequency of 110.7 was selected, but no signal was received. We had good ground contact, but we could not see the airport. It was hazy and as we descended the haze seemed to increase. Approach control told us to fly a heading of 220 degrees to intercept the localizer. The ILS was flagged. The autopilot was off. I called the controller and said, "Frankfurt approach, this is a different call sign, and I'm not receiving the ILS for runway 25L." The controller responded by informing us that the "frequency was now 110.3, tune your radio to 110.3 now." We switched to 110.3 and the ILS came in. As we were going through the localizer, we were given a heading of 270 degrees to intercept. We received the localizer and then saw the runway. We were told to contact the tower on 118.6. The Captain then said, "There is something wrong. This isn't Frankfurt." The Captain told me to ask the tower if they had us in sight. I called, "Frankfurt Tower, this is a different call sign, do you have a visual contact on us?" The tower responded with, "Affirmative. You are cleared to land." But again the tower used a different call sign. The Captain told me to call them again. I said, "Frankfurt Tower, this is an aircraft with a different call sign, do you have visual contact with us?" They again told us that they had us in sight and we were cleared to land. The Captain stated, "This is not the Frankfurt Airport. I'm going around." I said, "This is Frankfurt, there is runway 25L and there's the terminal on the right." I made this statement because I had been addressing the tower as Frankfurt Tower and there was no contradiction, and because I saw runway 25L painted on the runway. The Second Officer responded with, "It looks good," meaning the configuration of the airplane was set up for landing. We landed without incident. The ground controller asked, "Why have you landed here? Was this your alternate?"

Now look back through this story told by the First Officer and you will see points where clues were missed and situational awareness was lost.

1. The controllers began using a different call sign with the airplane, but the flight crew never cleared up this discrepancy. The First Officer was still trying to resolve the mystery of the different call sign when they were on short final at Brussels. The first time the problem came up was back with London Control.

2. Everyone in the crew was concerned about the rapid descent, "We commented among ourselves that we were getting down very early for Frankfurt." Brussels and Frankfurt are approximately 150 miles apart. When descending from FL370, a 150-mile closer target would produce a much steeper descent. But even though the crew talked about this unusually steep approach, they never questioned it. The airplane also had too much fuel remaining to have been at Frankfurt already. These were all clues to awareness that the crew ignored.

3. Recordings revealed that the Brussels air traffic controllers only said "Brussels" one time and just said "Approach" or "Tower" the rest of the time. This particular airline company did not have a scheduled flight to Brussels, and the controllers did not know why the airplane was coming there. They could have assumed there was trouble aboard the airplane, like a hijacking, or mechanical problems, but none of this was ever discussed.

4. The controllers asked the crew to fly to the Bruno VOR, but the crew could not find the Bruno VOR in the vicinity of Frankfurt. Here was a great opportunity to regain situational awareness. The instruction to fly to Bruno was part of the

maneuvering needed to get into position to land, but Bruno was not near the airport they thought they were flying toward. Instead of asking, they simply took a vector in the direction of Bruno. Then the second officer found Bruno near Brussels, not Frankfurt, but this fact was never acknowledged and they kept going.

5. The second officer could not raise the Frankfurt ATIS recording. She diverted much of her attention to this problem. The airplane was now descending between 8000 and 4000 feet and in reality Frankfurt was more than 150 miles away. The reason the recording could not be heard was more than likely that they were now too low and too far away to receive it. The steep descent, too much fuel, and the missing ATIS were all clues that they had not traveled far enough to be at Frankfurt yet, but nobody noticed. To rationalize the lack of ATIS, the second officer eventually concluded that the number 2 communications radio must be inoperative without considering the fact that they might have been out of range.

6. Both Frankfurt and Brussels have a runway 25L ILS approach. This coincidence set up the final approach mistake. The 25L approach at Frankfurt and the 25L approach at Brussels are very similar. They both have VORs north of the localizer, the one at Brussels being the Bruno VOR. Maybe the crew figured that the name of the VOR north of the Frankfurt localizer had been recently changed to Bruno. But nobody in the crew ever identified the Morse code for any station. When the off flags were showing on the 25L approach, the First Officer called approach and received a different

frequency. They tuned in that frequency and the off flags disappeared, but no positive identification was ever made.

7. When you believe things are one way, it is hard to accept evidence that things are actually another way. The First Officer thought he was approaching Frankfurt, so when the captain said, "There is something wrong; this is not Frankfurt," he could not believe it. He said, "This is Frankfurt, there is runway 25L and there's the terminal on the right." This crew had flown to Frankfurt before and the captain knew what he was seeing was not Frankfurt, but when the truth of the situation became unmistakably clear, their minds were not prepared for the jolt back to reality. They did not want to believe that they could have become so unaware of the situation that they could have landed not only at the wrong airport but actually in the wrong country! But there they were in Belgium, not Germany.

There was no accident involved with this flight so the normal postaccident procedures were not followed. The cockpit voice recorder tape was not removed and consequently the conversations of the crew as they approached the wrong airport were recorded over. The tapes of conversations between the crew and London ATC and Shannon, Ireland ATC were also lost. The captain and First Officer were fired by the airline. The second officer was taken off the flight line and sent back for additional training.

Clue after clue came and went, but each was ignored or explained away. As in so many other situations, the opportunities to regain situational awareness were there but they were not understood in time. To maintain situational awareness pilots must be open to more options and explanations. They must think that it is possible even for them to be "out of the loop" and not know it.

How can you stay "in the loop"? Are there clues to watch for that will be telltale signs that situational awareness is slipping away? Read through the "Keys to Maintaining Situational Awareness" and then look for these on your next flight.

Keys to Maintaining Situational Awareness

- "Wherever you are—be there!"
- Routinely do the routine things.
- Project the flight.
- Protect critical commodities.
- Avoid the self-inflicted wound.
- Make situational assessments.
- Prioritize your problems.
- Get to know your aircraft systems.
- Have a plan for distractions.

"Wherever you are—be there!"

This is a direct quote from Mr. Joel Smith, who is a training coordinator for the Boeing 747 fleet of Northwest Airlines. Joel is a former student of mine, and he invited me to sit in on one of his CRM training sessions one time. He had a roomful of veteran captains and first officers and nobody's attention wandered during his innovative seminar. I always say that I learn more from my students than they learn from me, and that day it was certainly true. Joel said a number of things that day that I have since used, but the most important thing he said was the simplest. He said, "Wherever you are—be there!" Which means that when you climb into an airplane, whether it be a 747 or Piper Cub, you must channel all your attention, energy, and resources to the situation at hand. It means that you must place yourself in the proper frame of mind and that you will not let self-distractions interfere with the flight. This is the first line of defense toward maintaining situational awareness. I say this phrase before every flight now.

Routinely do the routine things

You learned to use checklists when you began to fly, but some take the attitude that procedures and checklists are only for beginners. They think that checklists are great when you are first becoming familiar with the airplane, but with experience you can skip some steps. This attitude is not an "experienced" attitude, but a "complacent" attitude. The quickest way to lose awareness is not checking the routine items. Listen to the words of this pilot who landed gear-up because he did not do the routine things routinely.

ASRS NUMBER 447611 The approach was completely normal with no unusual circumstances or distractions. The only remarkable point was that on running the checklist, as printed on the panel behind the yoke, I held off on the landing gear because the aircraft was about 7 knots faster than the Vlo (landing extension speed) with 10 degrees of flaps. Without exception (but one) I check my three green lights at about 50 feet, but today I did not. I had a new set of tires put on the aircraft a couple of days before that had a higher profile than the previous ones. I was at the 50 foot point, concentrating on the pending flair and neglected to double-check the three green lights. No gear warning horn or light sounded and later it was discovered that these warning systems were inoperative.

They were inoperative due to a previously incorrect installation during a repair. Other than the pilot flying's ego, there were no injuries of any kind sustained. There was one passenger on board and we both evacuated to the side of the runway immediately after touchdown.

The pilot stated that he was running the checklist, but he "held off" on the landing gear because he was going too fast. This raises so many questions: why was he going so fast at that point in the checklist? Why didn't he just reduce speed? Why were the flaps down before the landing gear? In light airplanes we use the flaps for drag and slower/steeper approaches, but if drag is needed, the landing gear should be used first because it is mandatory. Flaps are optional. The gear warning horn was inoperative, but that would have been discovered during a proper preflight inspection.

There were so many departures from routine in this short story. When it comes to maintaining awareness, don't reinvent the wheel. Get back to basics. Use checklists. Follow standard procedures and read through the airplane's information manuals periodically.

Project the flight

To project the flight is to anticipate the flight. If you expected things to happen, you could never get caught unaware. Pilots use the phrase "staying ahead of the airplane." This means that your mind is out in front of the flight. In other words, when the airplane arrives at a certain point, the mind has already been there. The pilot sees the future of the flight and is making plans for the arrival of the actual airplane. Seeing the future is hard to do, however. Here are some strategies for "staying ahead."

A. Put Money in the Bank. Conventional wisdom about everyday life is that you should have a little savings set aside for a rainy day. In this way when your car breaks down you can get the car fixed and still pay the rent. The savings becomes a cushion or a buffer against the unknown or unexpected. To the pilot, putting money in the bank means doing all the extra things early. This preparation will act as a buffer against the unknown or unexpected flight event. If you are flying level en route and not much is happening at the time, look on the chart and see what frequency is likely to be your next hand-off. Dial that frequency into your second radio so it is ready to go. If something unexpected should come up (turbulence, traffic conflict, sick passenger, rough engine, equipment malfunction, etc.) before or during the controller hand-off, you will have less to do because the frequency will already be waiting on you and you will have more time to deal with the unexpected problem. You will be very busy approaching a high-density airport, so always have the next frequency ready. This way when the controller points out three other airplanes in your area, asks you to keep your speed up, make a tight base turn, follow the 737, report the airport in sight, and contact the tower, you can throw one switch on the audio panel rather than fumbling for the frequency and dialing it in.

Some VOR approaches do not require DME, but nevertheless have DME positions on the approach chart. Put money in the bank on the VOR approach by setting up the DME early and using it throughout the approach. You can determine your outbound distance, for instance, so that you do not fly outside the 10-mile "protected area" of the approach. On the way inbound you can get a better idea of how quickly you must descend in order to get down to MDA in time to make the airport. If you do not have DME, you could use a crossing radial from another VOR to mark the outside of the 10-mile protected area and an intermediate point along the final descent when you should be at MDA. There can be countless more examples of putting money in the bank. In each case we are doing something when we have time to better protect us for when we do not have time. There are times during any flight that can be downright boring. But the aware pilot knows that these times will not last. Soon the situation will change and the pilot workload will go from boredom to blowout. The next time you are flying along without much happening, remember to use this time wisely. Think of ways that you can put money in the bank.

B. Expect to Make Decisions. Understand that making decisions is as common to a flight as takeoff and landing. There are situations in flight when a decision is called for and every second that passes without making that decision makes the situation more critical. In these situations some pilots are just unaware that a decision is staring them in the face and screaming "solve me!" Understand that a continuous stream of decisions is normal and necessary to the safe outcome of the flight. Some pilots take the attitude that controllers will not let them get into too much trouble, or things will just work out because things always seem to work out. The accident reports make it clear that these attitudes lead to crashes.

The solution? Pilots must realize that making decisions is their job as pilot in command and that there will be countless points of decision on every flight. Pilots must accept this fact and start to expect to make decisions because decision making is normal. When flying along to a destination and when the pilot workload is low, the pilot should be planning for and anticipating the decisions that are bound to present themselves. Which runway is in use at the destination? How should I maneuver to be in a position to enter the pattern? Should I expect to fly an instrument approach at the destination? If I do shoot the approach, should I be planning on a straight-in or circling approach? These are examples of decisions that are inevitable. You will not be able to see perfectly into the future, but you can plan and expect to make these decisions as a routine function of any flight.

Decisions are not the product of some abnormal flight situation that a pilot might never face in a career of flying; they are everyday, normal, routine. Pilots should expect to make decisions on every flight. Pilots must search, like detectives, routing out the hidden decision needs that call for PIC attention.

C. Make Contingency Plans. Pilots must never waste any time. When expert pilots fly, they are doing something even when the workload is low. They may have only been holding an altitude and tracking a VOR radial, but they are always thinking and doing something. During these times the pilots should play little mind games. They should play "What if?" Ask yourself as you fly along, "What if I started to get a rough-running engine, what would I do? Are there airports nearby that I could divert to? Are there fields below I could land in? The engine will probably not run rough and the flight will probably continue on without any problems, but asking "what if" makes you think of a contingency plan just in case. Have you ever heard the saying, "Always leave yourself an out?" Leaving yourself an "out" means having a backup plan. You should switch to the backup plan anytime the first plan does not seem like a good idea anymore. But you must have a backup plan waiting in reserve in order to be able to switch. Playing "what if" forces you to think up a backup plan.

What if I get to the decision height and cannot see the runway? What if I am told to switch frequencies by ATC but cannot get anyone to answer on the new frequency? What if my approach clearance gets canceled right here and I am asked to do a holding pattern? What if the glide slope goes out in the middle of an ILS approach? What if I start picking up ice at this altitude? Each of these "what if" questions when asked would force the pilot to answer with a contingency plan. Someday one of your "what if" questions might come true, but because you have asked what if and prepared a response you have a ready-to-go contingency plan. On that day you will not waste valuable time wondering what you should do. You will quickly decide to switch plans, solve problems, and move on to the next "what if."

D. "The next important point will be…" Practice anticipation during the low workload times by filling in the blank: The next important point will be_____. I have seen this situation played out many times. A controller asks a pilot to fly to a particular fix like a VOR or an NDB and hold over the station. The pilot turns the airplane and begins to fly towards the station. There may be a crosswind, so the pilot works out an angle that will offset the wind and create a straight course to the fix. Then we fly in silence for the several minutes it takes to get to the fix. All of a sudden, as if it's a big surprise, the station is passed and the pilot now does not know what to do in order to enter the holding pattern. What kind of hold entry should I use? Do I turn? If so, what heading should I turn to? Which direction should I turn? All of these questions come into the pilot's head in a mad rush as the station is passed. Confusion sets in, altitude is lost, radio communications are forgotten, the holding pattern's protected area is flown out of, in short, all hell breaks loose. All that could have easily been avoided if the pilot had used the time traveling to the fix to plan for what was needed after the fix. So to help pilots remember to plan ahead, fill in the blank: The next important point will be_____. That point might be passing a holding fix and that will remind you to plan for the hold entry to follow the fix. The next important point may be an intercept and a descent on an approach. It might be reporting your position flying inbound to a VFR tower. It could be anything, but ask yourself what it is and you will be planning ahead.

Projecting the flight then means using time wisely, putting money in the bank, expecting to be in decision situations, asking "what if" questions, and reminding yourself what important event is coming up next. If pilots do this they will be better prepared when and if things go wrong. They will be less likely to be caught off guard or to be unaware when something is going on that they should know about. This all implies, of course, the pilot must keep an active mind during the entire flight. There is no room for the pilot to doze off mentally or to be thinking about nonflight problems. Mental alertness leads to situation awareness.

Protect critical commodities

When you fly you must maintain control over your two critical commodities: time and fuel. If you lose track of these two, then you have lost your situational awareness—it's as simple as that! Before any flight you must have a time and fuel plan, and then you must stick to and monitor that plan. The following accident report is a story of a pilot who lost awareness of time and fuel.

NTSB NUMBER MIA90FA106   The pilot departed Naples Municipal Airport, in Naples, Florida, VFR on a personal flight at 11:30 hours. The airplane was observed conducting touch and go landings at the Marco Island Municipal Airport in Marco Island, Florida between 1300 and 1400 hours and then departed at an unknown time. The airplane wreckage was initially spotted at 1700 hours in a bay, but was thought to be a sunken boat. The Civil Air Patrol spotted the airplane at 1925 hours and notified the authorities, who responded to the crash site. Aircraft records and Hobbs meter in the airplane, indicate it had been operated 4 hours and 36 minutes. Aircraft performance charts indicate fuel exhaustion at 4 hours and 24 minutes. There was no evidence of fuel at the crash site or purchase of fuel by the pilot. Physical evidence indicates inflight loss of control. The airplane stalled and spun to the left and impacted the water. There was no evidence of any powerplant, airframe, or flight control failure.

Probable Cause: Fuel exhaustion due to improper planning and decision by the pilot, and an inadvertent stall/spin while maneuvering for a forced landing.

The probable cause attributes this accident to "improper planning" before the flight and "improper ...decision" during the flight. The pilot did not have a time and fuel plan and did not monitor his precious commodities. The next report was written by a pilot who also did a poor job of fuel/commodity management.

ASRS NUMBER 409296 A weather briefing from Altoona flight service indicated marginal VFR during the first part of my VFR flight with improving conditions in the last 1/2 of the trip (after Zanesville, Ohio). Current weather showed stations better than MVFR with possible 2 to 7 miles in fog and mist on the first part of the flight. The trouble spot extended from Wheeling, West Virginia through Zanesville. I calculated groundspeed at 90.3 knots with a true airspeed of 105 knots. Total time en route was estimated at 3 hours and 23 minutes over a distance of 306 miles. Total fuel aboard was estimated at 4 hours and 30 minutes. The problem arose en route when I had to take a more westerly course to remain in VFR weather. I stayed west of Zanesville and continued back to my original course south of Zanesville. I lost about 30 minutes by circumventing the problem area. As I got closer to Lexington, I was uncomfortable with the fuel situation, however, I was within the time frame of my fuel estimate. Fifty miles northeast of Lexington, I decided to refuel at FGX. I circled the field and landed but was informed that no fuel was available (the tanks were being replaced). I received this information while landing and decided not to shut down the engine, but to continue on to Lexington. I calculated total flight time would be 4 hours an 10 minutes. Four hours and 5 minutes into the flight, the engine stopped 6 miles short of Lexington airport. I was in contact with Lexington approach and radioed that I was out of fuel and landing in a pasture field below. When on the ground, I used a cell phone to talk to Lexington airport and arrange for fuel to be transported. We estimated the distance of the field and the takeoff performance of the airplane. I loaded 10 gallons of fuel and made a soft field takeoff with no incident and then landed at Lexington Blue Grass airport. In looking back at the situation, I could have avoided this problem with better fuel management inflight and/or by setting a personnel safety factor for reserve fuel. In the future a one-hour reserve plus better fuel management will prevent this from happening again.

The challenge for any pilot is to manage the resources they have (fuel and time) and end the flight safely on the ground with some fuel and time left over. The challenge of time and fuel management is not restricted to small aircraft. An air carrier crew flying to Mexico had a difficult time/fuel problem that was compounded by a language barrier.

ASRS NUMBER 428510 Upon arriving in the terminal area, the weather had deteriorated to Instrument Meteorological Conditions (IMC) with thunderstorms in all quadrants, heavy precipitation, rime ice, and lightning encountered in the descent. Approach control advised us to expect the ILS runway 23L approach. They reported heavy rain at the airport with windshear. We advised the controllers that we were breaking off the approach and requesting vectors to the east to a point where we could hold until the weather improved. Approach control complied, we entered a holding pattern east of Otumba on the 110 degree radial in clear weather. After holding for several minutes, with no improvement in the weather at the airport, and then with deteriorating weather at our holding position, we elected to divert to our alternate, Acapulco, for fuel. After several very frustrating minutes trying to communicate our desires, approach cleared us direct to TEQ. But that route ran right through the weather. We advised ATC and requested radar vectors to the south of our position. Again numerous attempts were needed to communicate our desires. Finally we told ATC that fuel was an issue and that we were requesting radar vectors direct to Acapulco. They finally complied and we proceeded to Acapulco and made an uneventful visual approach and landing in Visual Meteorological Conditions (VMC). A major issue was the language barrier. We were the only English speaking aircraft of the frequency. Communication became difficult as we were asking for things that are not normal for arrivals into Mexico. Also it was difficult to understand what other aircraft were doing. ATC seemed to finally understand when we stated our fuel condition.

The combination of little time, low fuel, and a language barrier eventually worked out in this story, but one day near New York the same combination of factors proved to be fatal.

NTSB NUMBER DCA90MA019   On January 25, 1990 at approximately 2134 eastern standard time, Avianca Airlines flight 52, a Boeing 707-321B with Colombian registration HK-2016, crashed in a wooded residential area in Cove Neck, Long Island, New York. Avianca 52 was a scheduled international passenger flight from Bogotá, Colombia, to the John F. Kennedy International Airport, with an intermediate stop at José María Córdova Airport, near Medellín, Colombia. Of the 158 persons aboard, 73 were fatally injured. Because of poor weather conditions in the northeastern part of the United States, the flightcrew was placed in holding three times by air traffic control for a total of 1 hour and 17 minutes. During the third period of holding, the flightcrew reported that the aircraft could not hold for longer than 5 minutes, that it was running out of fuel, and that it could not reach its alternate airport, Boston's Logan International Airport. Subsequently, the flightcrew executed a missed approach to JFK Airport. While trying to return to the airport, the aircraft experienced a loss of power to all 4 engines and crashed approximately 16 miles from the airport.

Probable Cause: The failure of the flightcrew to adequately manage the airplane's fuel load, and their failure to communicate an emergency fuel situation to air traffic controller before fuel exhaustion occurred. Contributing to the accident was the flightcrew's failure to use an airline operational control dispatch system to assist them during the international flight into a high-density airport in poor weather. Also contributing to the accident was inadequate traffic flow management by the FAA and the lack of standardized understandable terminology for pilots and controllers for minimum and emergency fuel states. The Safety Board also determines that windshear, crew fatigue, and stress were factors that led to the unsuccessful completion of the first approach and thus contributed to the accident.

Losing track of time and fuel is equivalent to losing situational awareness. Keep a navigation log and compare your actual time to your estimated time on each flight. Monitoring the amount of time you are in the air and therefore the amount of fuel you have used is a vital way to maintain awareness.

Avoid the self-inflicted wound

Pilots have enough problems in the dynamic flight environment without making things worse on themselves. But pilots have imposed unnecessary distractions on themselves that compounded and aggravated an already challenging situation.

ASRS NUMBER 443877 The copilot was flying and was cleared for the ILS runway 27R approach into Philadelphia. After we were cleared to 2,100 feet MSL, the copilot was trying to get the autopilot to lock on for a coupled ILS to runway 27R while turning right and passing the Speez Outer

Marker. During the landing checklist the airplane started descending through 2,100 feet after passing Speez at around 1,500 feet per minute. After I mentioned our altitude was low, being 2 dots below the glide slope, he started leveling off at 1,200 feet. Then the tower asked us to check our altitude because they were getting a low altitude alert. The copilot maintained 1,200 feet until the glide slope was recaptured and continued on to an uneventful landing. It appeared that we lost track of the rate of descent while looking at the ground for a visual reference to our position. Also, the copilot's attempt to lock-on with the autopilot failed for some reason and this provided an unneeded distraction. Our visibility straight ahead was approximately 2 statute miles, but we could see the ground clearly. I'm not sure if he was looking at the airspeed bug and trying to slow the aircraft or not, but his scan failed to see the low deviation from the glide slope. Doing the landing checklist caused me to not catch our low deviation until 1,500 feet MSL, at which time corrections were put into place to level off. I think a little earlier preparation on base leg could have helped us from being rushed on final. The copilot should have used raw data to manually fly the approach instead of wasting precious time trying to get the autopilot to capture the approach. Once he did that, the remainder of the approach and landing were fine.

Wasting time with the autopilot created a self-distraction that resulted in a dangerous low-altitude situation. An autopilot can be a great way to reduce the workload, but in this case it got in the way, imposed a pilot-induced distraction, and could have caused an accident.

ASRS NUMBER 441955 This was the Captain's first trip from Aviano. My second trip from Aviano, and the Flight Engineer was also unfamiliar. Our clearance was issued by ground control just prior to reaching the active departure runway. The clearance to the en route phase of flight was via the "Vincenza 6B departure." Our resources included the commercial charts (2 approach procedures and the airport chart). The government FLIP charts (same as commercial), government revisions (which did not have the Vincenza 6B SID), and a government airport book (which had lots of verbage on the airport, but no SID). At about the same time that we exhausted our onboard resources, I recalled that on my one previous trip from this airport, the Captain had pulled from his collection a photocopy of the VIC 6B departure—the source of which remains a mystery to me. I shared what I could remember from the departure procedure with my Captain. He had me advise the controller that we found the departure procedure. I was a bit leery. His idea worked, but the controller asked if we had the issue showing the right turn at 600 feet, etc. The controller apparently read to us the entire SID. I scribbled it all down, we set our navigation equipment, called for the takeoff clearance, and departed. Our mistake here was our failure to brief what we had just heard (because we had to make a time slot). After takeoff and the first two turns, the Captain acquired and then navigated direct to Venezia, which was about 30 nautical miles to the south (Vincenza and Venezia can sometimes sound the same!). I thought that I had heard "direct to Vincenza" as part of the SID. I figured that I must have misunderstood. After about two minutes and 8 to 10 nautical miles on the southbound course, the controller advised us to turn immediately toward, and fly to Vincenza, which was about 30 miles to the west of the airport. We did, and proceeded without further incident. Flying around without the necessary charts on board is not a good idea.

This was clearly a self-inflicted wound. The crew did not have the proper charts but proceeded anyway. The First Officer was "leery" but did not speak up.

ASRS NUMBER 451081 We departed runway 25 from Denver on a 255 degree heading as part of the Denver 3 Standard Instrument Departure procedure. Then we had Radar vectors to the Lufse Intersection. Our destination was Colorado Springs. The tower delayed our hand-off to departure control due to frequency congestion. Departure initially gave us a heading of 240 degrees and a climb to 12,000 feet MSL. We were then given a heading of 190 degrees to intercept Victor 81. We turned to the assigned heading, but with easterly winds aloft, we did not intercept the airway quickly enough. We were below the OROCA (off-route obstruction clearance altitude) and the controller received a terrain warning. Since it is only a 60-mile flight, I had already switched to the Colorado Springs ATIS frequency. When I returned to the Denver departure controller, we turned to a 100-degree heading as directed by the controller. I was unaware that there was a problem until we landed at Colorado Springs and the ground controller gave us an ATC number to call. Normally we depart Denver on runway 17 Left or Right, or runway 8 to go to Colorado Springs. However, this time we departed runway 25 towards terrain. I was unfamiliar with Victor 81 and had to look it up on the chart, which caused a delay in intercepting. This is a difficult leg because everything is happening very fast.

This short, 60-mile flight took all the elements of a longer trip and compressed it into just a few minutes. But rather than being more prepared, the pilot relied on past expectations. He was used to taking off away from the terrain, *"Normally we depart Denver on runway 17 Left or Right, or runway 8 to go to Colorado Springs,"* but this time they were doing something abnormal. The crew created a distraction for themselves by not becoming completely familiar with the new route before takeoff. The pilot said, *"I was unfamiliar with Victor 81 and had to look it up on the chart, which caused a delay in intercepting."* Then when the airplane flew over higher terrain and the controller's low-altitude alert went off, the crew was not on the frequency. They had turned away from approach control and were listening to ATIS when their potentially lifesaving instructions came from ATC. By switching frequencies at that time, they contributed to their own problem. It is one thing to have a problem and realize it; it's another thing when you have a problem but are so unaware of the situation that you learn of the problem from a ground controller after landing. *"I was unaware that there was a problem until we landed at Colorado Springs...."* This crew created self-distractions and got too close to the terrain for comfort.

Situational assessment

In order to maintain situational awareness, pilots and controllers must continually be making an evaluation of what is taking place. The evaluation must be followed with an assessment of condition. Is the condition favorable? Is everything working out as expected, or are concerns foreseen? It is only after a cool situational assessment is made that competent decisions can be made. After all, good decisions are made when the greatest number of facts are known. Assessing the situation is knowing and understanding the implication of the facts at hand. Read the accident report of the fatal accident that took place on what was to have been a short hop from Rome, Georgia, to Huntsville, Alabama.

NTSB NUMBER DCA92MA011 Before takeoff, an IFR flight plan was filed for a 15-minute flight from Rome, Georgia to Huntsville, Alabama (in a Beech 400 jet). Takeoff commenced at 0937 EST with the copilot flying the aircraft. After a VFR takeoff, the Captain contacted Atlanta Center to obtain an IFR clearance. The controller advised that other traffic was in the area and instructed the flight to remain in VFR while an IFR clearance was being arranged. At that time, the flight reported at 1,300 feet MSL in VFR conditions. While waiting for an IFR clearance, the crew became concerned about higher terrain and low ceilings. At about 0940, the Captain directed the copilot to fly "back to the right." Approximately 1 minute later, the cockpit voice recorder stopped recording and radio contact was lost with the aircraft. Later, the aircraft was found where it had collided with the top of Mount Lavendar. Elevation of the crash site was approximately 1,580 feet MSL. The aircraft was not equipped with a ground proximity warning system.

Probable Cause: The Captain's decision to initiate visual flight into an area of known mountainous terrain and low ceilings and the failure of the flightcrew to maintain awareness of their proximity to the terrain.

Of course we were not there, and there may have been unknown circumstances that led these pilots to their decisions. But it appears that the pilot could have made a better situational assessment that could have lead to a better outcome. Seeing that the flight was trapped between the ground and the clouds without being able to get a clearance right away, the crew could have just landed back at Rome and received a clearance void time. The "traffic in the area" was another airplane on the instrument approach so the crew could have flown away from the higher terrain, but making sure to stay clear of the instrument approach corridor. As you fly, do not become numb. You should continuously take a critical eye to the events that surround you. This will keep you alert and ready to make the next decision.

Prioritize your problems

Pilots are confronted with an inflow of information (radio communications, spotting other traffic, flight and engine instruments, etc.) at all times during a flight. We can only do so much with this information at one moment in time, so we must decide which bit of information or which potential conflict we should attack first, second, and third. We must accurately assess the array of information or potential conflicts that are present and elect to work on the most critical first. Pilots must learn to choose the most important or urgent item that is present and quickly turn their attention to that problem. Once that problem is solved or mitigated, we move on down the priority list. The following accident report, unfortunately, is the story of a pilot who did not prioritize.

NTSB NUMBER LAX91FA020   The aircraft (a Grumman AA-1B) was loaded over its gross weight with a center of gravity outside the rear envelope limit. The pilot in command and a media reporter proceeded to the site of an auto accident. Evidence supports the fact that the reporter, who was currently receiving dual instruction and held an expired student pilot certificate, was flying the aircraft. The aircraft violated many company minimum altitude limits by descending to about 800 feet above the ground and initiated a series of steep turns at slow airspeed. Statements from other pilots established that the reporter liked to cross control the aircraft in order to have a better view of the ground. Ground witnesses observed the aircraft perform an abrupt maneuver then spin to the ground.

Probable Cause: The inadvertent stall/spin. Factors in the accident were: The pilot's decision to allow the reporter-passenger to fly the aircraft in a critical flight situation, his decision to allow the aircraft to descend to an altitude too low to allow for contingencies, and the failure of the pilot in command to ensure that the aircraft was properly loaded.

The pilot was highly motivated to "get the story" of the auto accident, so highly motivated that he did not keep his priorities straight. The most important item should have been not to create an even greater media event by having an airplane crash at the scene of a car crash. The first priority should have been to fly the airplane at a safe altitude, with safe maneuvers, and to have the airplane safely loaded.

As you fly ask yourself what is the most important item that you can address right now. Line up the workload tasks in order of importance and work through the list based on priority. The mere fact that you are thinking about priorities will improve and maintain your awareness.

Maintain high technical proficiency

Remember the story of the pilot who landed off airport and struck power lines (FIG. 3-1)? Part of what drove him to make that fateful decision was his lack of systems knowledge. He thought he had no electricity because he misunderstood the airplane's ammeter. He thought his radios were broken because he did not remember that VHF radios only work with line of sight. Because he did not know how things worked, he was unable to troubleshoot. One misconception lead to another and soon he had painted himself into a corner where an off-airport landing looked like a good option.

One of the scariest situations a pilot can be placed in is when something is going wrong but the pilot doesn't know it yet. Worse still is a situation where the pilot knows something is wrong, but does not know what to do about it. The solution is to become knowledgeable about the airplane's systems. Being a renter pilot is no excuse. You must know how things work on any airplane you fly. I have known some pilots (usually young and with little experience) who have taken the attitude that learning about the airplane's systems is just not their job. "That's for maintenance guys!" they will say, "I'm a pilot; I don't need to know how to fix it." This is quite a naive attitude. Maybe a pilot will not be called upon to actually repair a system problem, but it will most likely be a pilot that discovers the problem—sometimes in midflight. When a problem is detected, what the pilot does next will affect the safety and outcome of the flight. We pilots might not be able to repair the problem, but we should know not to make it worse, and whatever we do, we must bring the aircraft back down safely. Pilots must be system smart.

I teach at a university that specializes in training airline pilots. The regional air carriers is their first step. When my former students are hired by one of these commuters, they are placed into a new hire class that can last as long as three weeks. During those weeks the new airline pilots are subjected to an intense barrage of classwork. They call it "like drinking from a fire hose." What do you think these classes spend the most time on? Weather? Navigation? Regulations? Aerodynamics? Instrument flight? No, they spend virtually no time on these topics—it was my job to teach that back in college. No, the course spends about one day on company policies and the rest is aircraft systems, systems, and more systems. The company figures that if they do not already know how to fly, understand weather, or navigate, they would not have even been hired, so the only thing that they have not been directly exposed to is whatever airplane they have been hired to fly. So the new hire class is really an airplane systems course. Why do these airlines spend so much time on systems training? Because they must be sure that their pilots will be system savvy enough to do the right thing at the right time.

Every pilot should do the same thing—take a systems course. Now there may not be an organized systems course at your home airport for a Cessna 152, or whatever you fly, but you do have manuals and maintenance technicians. Even an airplane as seemingly simple as a Cessna 152 has all the basic systems: electrical, ignition, vacuum, hydraulic (brakes), flight control, fuel, and pitot-static. Start with the manuals, but to be honest you will learn more in an hour from an A&P technician than 10 hours reading the book. Get to know your technicians.

Some shops have insurance rules against your being in the shop when work is being conducted, but at most places the technicians will be happy to have you look over their shoulder. The best idea is to go with your airplane into one of its routine inspections. As a result you will become a much better systems troubleshooter, and this means better situational awareness.

Have a plan for distractions

Every flight has minor distractions, and someday you probably will face a major distraction. Since distractions are inevitable, it pays to have a ready-made plan of attack to handle the distraction. First, no matter what the distraction is—from an engine fire to a stuck air vent—you must continuously fly the airplane. Remember the Eastern Airlines flight that crashed in the Florida Everglades. All three members of that crew became distracted by a burned-out nosewheel light. Somewhere in the sequence of events the autopilot became disengaged and the airplane started a slow descent. It was a dark night and over the Everglades there are no ground lights, so the pilots never saw it coming. The crew tried to remove the old light, but it would not come loose. They got a handkerchief for a better grip, but that didn't work. They continued discussing only the light until at 50 feet AGL, one of the pilots said, "Hey, something is wrong with the altitude; we're still at 2000 right." This was followed by the sound of impact. Even though there was three pilots in the cockpit, nobody was flying the airplane.

Most stall-spin accidents are the result of a distraction. Airplanes are not just cruising along and all of a sudden they enter a spin. No, spins are the result of a stall that was entered into by an inattentive pilot. Investigators find that spin accident aircraft have a fuel selector valve incorrectly placed, or have one full tank and one empty tank. The pilot became distracted but did not have a plan. They did not remember that in every situation, flying the airplane is priority one.

The next step in the distraction plan should be an assessment of the problem, but this cannot take long and may be obvious. Once the problem has been identified, the next step is to troubleshoot using a division of attention. Dividing attention means that you will not work on any one item very long. Pilots tend to prefer working on projects until they are completed, but in the airplane you often must leave a job unfinished, shift to something else, and then come back to the job. If you must get an instrument chart out of a book, you must not forget to fly the airplane while you flip through the pages. You cannot set up and identify two communication and two navigation radios, a DME, an ADF, and test the marker beacons in an uninterrupted sequence. That sequence would take too long and the airplane will demand attention in the meantime. Divide up the jobs, moving your attention back and forth. Never fix on a single instrument, task, or problem. Pilots must juggle everything, holding nothing still for very long. As you work through and resolve the problems, the workload will reduce and the challenge of the distraction will pass.

So be ready and expect to use a distraction plan. Fly the aircraft. Assess the situation. Divide your attention. Resolve the distraction. This is the best strategy for maintaining safety and awareness.

The Best Example of Maintaining Situational Awareness

The flightcrew of a regional air carrier was faced with a very unusual problem one day. Their landing gear came down, but when it did the nosewheel was cocked to one side and not aligned with the direction of aircraft travel. How the crew worked through the problem is a wonderful example of working together, knowing the airplane and its systems, remaining cool under pressure, and above all maintaining situational awareness.

ASRS NUMBER 447168   During the VOR runway 6 approach to Martha's Vineyard, the crew observed an unsafe gear indication. The nose gear light was not illuminated. The First Officer flew the aircraft and the Captain told the tower we needed to go missed approach and hold to figure out what the problem was. The crew chose to do a "fly-by" to have the tower take a look to see if the gear was down. The Martha's Vineyard tower advised that the gear was down but at a 70 to 80 degree angle. The crew executed the missed approach and received delay vectors from the cape approach control. The Captain raised the gear handle and performed the manual gear extension checklist: Both circuit breakers passed, but warning horns failed and warning lights failed. The gear was then manually extended by the Captain. All three landing gear lights came on, and the gear unsafe handle light was off. We then began another VOR runway 6 approach into Martha's Vineyard. On final approach the tower advised us that the nose gear appeared to be at a 90-degree angle, and offered to allow us to circle the traffic pattern while we decided what to do. Martha's Vineyard tower then closed the airport, and advised us that the airport was "ours" and that there would be equipment standing by. While remaining in the traffic pattern, the First Officer flew the aircraft and the Captain spoke with officials of airline company. The aircraft was flown at 700 feet MSL while circling. The tower then asked if we could give them about 5 minutes to allow additional crash-fire-rescue equip-ment from around the island to arrive at the airport. The tower also advised us that the state police had called inform-ing them that there was limited hospital facilities on the island. We had 9 passengers. On the company frequency, personnel from the operations department asked us to consider diverting to Boston and the need to burn off some fuel. The Captain determined that one hour of fuel should be used up before a landing attempt. During our circling we considered fuel burn, hospital facilities, crash-fire-rescue equipment availability, runway length, and terrain immediately adjacent to the runway and overrun area. The two main factors that influenced the decision to divert to Boston instead of landing at Martha's Vineyard were that if we were going to burn one hour of fuel anyway, we might as well go back to Boston where there is a larger runway and more capacity at the hospitals. While en route to Boston we briefed the approach and rebriefed the evacuation and engine shutdown procedures. Boston approach gave us delay vectors in the shape of a box when we arrived about 15 miles south of the airport. When the fuel on board reached approximately 300 pounds, we began the ILS runway 4R planning on a side-step maneuver to runway 4L. We decided to land with flaps set at 35 degrees, and hold the nose off as long as possible. It was mentioned from the company chief pilot during conversations on the company frequency, that one method that had been discussed was to come in with reduced flaps and a higher speed. Then at touchdown try to skip the nose wheel on the runway to try and straighten out the nose gear. We chose not to try that. The tower switched us to a separate frequency of our own with only the controller, crash-fire-rescue people, and us. We broke out of the clouds at approximately 1,500 feet MSL and began the side-step to runway 4L. We touched down approximately 10 knots fast, with the airplane at 13,000 pounds. The Captain held the nose wheel of the runway, allowing it to first touchdown at about 40 knots. When the nose wheel touched down, the nose of the aircraft initially jerked to the right, but no more than about one foot. The Captain then lifted the nose and then allowed it to touchdown a second time. It seemed to skip once and then straighten out. The Captain used full left rudder and differential braking to maintain directional control. The Captain allowed the aircraft to roll with minimum braking to stop the aircraft at taxiway N1. During the rollout the passengers applauded. After bringing the aircraft to a stop, the Captain secured the engines using memory items and the First Officer advised the tower that we were evacuating the aircraft on the runway. We chose not to taxi the aircraft off the runway because we did not know the condition of the landing gear. We were concerned that if the gear collapsed while taxing, the prop blades would strike the ground, shatter and come through the fuselage. The Captain left the cockpit first and opened the door while the First Officer finished talking to the tower and retrieved the fire extinguisher for under his seat. The Captain and First Officer exited the aircraft and the Fire Chief asked how many people were on board. I told him 9. The First Officer then reentered the aircraft and asked the passengers to exit and follow the instructions of the Fire Chief. The First Officer apologized to the passengers for the inconvenience and made the comment, "I couldn't imagine what they went through back there." The passengers thanked the crew and told them they had done a great job.

Look back at some of the steps that this flight crew took to maintain their awareness and make it possible to make lifesaving decisions. At the first hint of a problem, the crew asked the Martha's Vineyard tower for a missed approach and a "fly-by." This bought the crew some time to assess the situation and start to formulate a plan. The crew effectively divided up the cockpit duties. The First Officer flew circles around Martha's Vineyard airport while the captain went to work on the landing gear. The crew was technically competent. They wasted no time running the proper checklists and they were able to properly diagnose the problem because they knew the airplane's systems. They sought outside help, not only from the controllers, but also from their own company officials. Then they took into consideration all the factors involved: "During our circling we considered fuel burn, hospital facilities, crash-fire-rescue equipment availability, runway length, and terrain immediately adjacent to the runway and overrun area." These factors were taken together and they arrived at a "best plan." The captain always stayed in command of the flight. Even when his boss, the airline's chief pilot, gave him a suggestion on how to attack the cocked nosewheel, he trusted his own intuition. And finally, this crew really flew their airplane well. They displayed excellent "crew resource management," but they also had excellent flying skills. The captain held the nosewheel off during the rollout, then touched the nosewheel down in an attempt to straighten it out. He brought the nose off the ground and set it back down while using full rudder, and opposite braking. That was some fine flying. Some say that CRM and flying skills are at opposite ends of a continuum and that in recent years too much attention was spent on CRM and not enough on flying skills. This crew blended both. At no time did the situation get away from them and carry them away. They were in command, aware, and that led to their safety. The passengers said it best: "They did a great job."

4

Recognizing the Loss of Situational Awareness

Have you ever said to yourself, "If I knew then what I know now...." This statement is the recognition that at one point in time you did not know the whole story. You did not know all the facts, and as a result you decided on a course of action that you might not have selected had you known all the facts. This is also a recognition that at one point in time you had lost (or never gained) situational awareness. We know that in an airplane the loss of the big picture can lead to accidents. We also know that the loss of situational awareness is at times inevitable. So when situational awareness is lost, it is the pilot's job to recognize the loss and take immediate steps to regain the big picture.

A tragic example of what can happen when pilots and controllers lose the big picture took place in Sarasota, Florida, when two Cessnas collided on the ground during their takeoff runs.

NTSB NUMBER MIA00FA103A; SARASOTA, FLORIDA   On March 9, 2000, about 1035 Eastern Standard Time, a Cessna 172K on a personal flight and a Cessna 152 on an instructional flight collided during takeoff on runway 14 at the Sarasota-Bradenton International Airport, in Sarasota, Florida. Visual meteorological conditions prevailed at the time and no flight plan was filed for either flight. Both aircraft were destroyed. The airline transport-rated pilot and another pilot in the Cessna 172, and both the flight instructor and student pilot in the Cessna 152, all were fatally injured. Both flights were originating at the time of the accident. Air traffic controllers at the FAA Sarasota-Bradenton International Airport Air Traffic Control Tower stated that the flight instructor of the Cessna 152 called for taxi instructions from the Dolphin Aviation ramp and was told by the ground controller to taxi to the end of runway 14. The ground controller then went on a rest break and a supervisor took over the ground control position. Draft transcripts of communications show that about 2 minutes after the Cessna 152 called, the pilot of the Cessna 172 called for taxi instructions from the Jones Aviation ramp and was told by the supervisor ground controller to taxi to runway 14. [*The Dolphin and Jones FBO ramps are on opposite sides of runway 14.*] The supervisor stated that he thought the Cessna 172 was leaving from the Dolphin Aviation ramp and would taxi to the end of runway 14 and not the intersection of runway 14 and the foxtrot taxiway, which leads from the Jones ramp. The paper strip *[air traffic controllers record of the movements of the Cessna 172]* was marked for the end of runway 14. Draft transcripts show that the pilot of the Cessna 152 called ready for takeoff at 10:30:46. The local controller told the pilot to hold short of the runway. At 10:32:51, the pilot of the Cessna 172 called the local controller stating he was the number 2 aircraft ready for takeoff. The pilot was told to hold short of the runway. At 10:34:00, the pilot of the Cessna 152 was told by the local controller to taxi into position and hold. The local controller then cleared a third aircraft into position and hold at the intersection of runway 14 and the foxtrot taxiway. That aircraft was then cleared for takeoff. At 10:34:54, the local controller cleared the Cessna 152 for takeoff on runway 14. The local controller stated that he then saw the paper strip for the Cessna 172, which showed that the aircraft was at the end of runway 14, looked out and saw a high wing Cessna in the first position at the end of the runway, and thought it was the 172. At 10:34:57, he then cleared the 172 into position and hold on runway 14. He then diverted his attention inside the tower cab, and a short time later he observed the fire which resulted from the collision of the Cessna 172 and Cessna 152. The sound of an emergency locator transmitter was heard at 10:35:13. A witness stated he observed the 152 begin a takeoff roll from the end of runway 14. At about the point where the aircraft obtained takeoff speed, the Cessna 172 entered the runway from a taxiway on the left side of the runway [*taxiway foxtrot*]. The Cessna 152 lifted off and got a few feet in the air and initiated a right turn, in what appeared to be an attempt to avoid collision with the 172. The Cessna 152 then appeared to stall and crash into the 172. Both aircraft immediately erupted in flames and came to rest on the runway together.

The supervisor ground controller entered the picture halfway into the sequence and believed that the Cessna 172 was behind the Cessna 152 on the same taxiway. The controller did not put taxiway foxtrot on the information strip, as is standard procedure. Maybe if the controller had written down foxtrot on the strip, it would have alerted him to the error. Information strips on previous airplanes departing Jones on taxiway foxtrot that day did have the notation: "foxtrot." Forgetting to write down the taxiway on the strip did not by itself cause this accident, but it was one link in the chain.

There were three airplanes involved. Two were at the intersection of foxtrot and runway 14 and one was at the end of runway 14. But the controller had it backwards. He thought one was at the intersection and two were at the end. At 10:32:51, the pilot of the Cessna 172, who was actually number 2 at the intersection, called the local controller, stating he was "the number 2 aircraft ready for takeoff." The controller must have thought he meant number 2 behind the Cessna 152 at the end of runway 14. The Cessna 172 pilot was not as clear and not as aware as he should have been here. The Cessna 172 pilot should have noticed that another airplane, the Cessna 152, was also preparing for takeoff down at the end of the runway. If he had seen the other airplane, he could have made a better radio call by saying, "Ready for takeoff at the intersection of runway 14" or "Ready for takeoff runway 14 at foxtrot." Either way this would have clarified the Cessna 172's exact position. The 172 pilot, however, made a vague statement that allowed the controller to continue to believe the wrong position. The first airplane at the intersection was given permission to taxi into position and hold, and the Cessna 152 was given position and hold behind him. Those two airplanes were now staggered in position on runway 14. The lead airplane was given the takeoff clearance and away they went. The Cessna 152 was given a takeoff clearance and began the takeoff roll. The controller then, still thinking that the 172 was behind the 152, gave the 172 the clearance to roll onto the runway and hold. But the 172, being at the intersection, not the end, rolled in front of, not behind, the 152. The controller had an excellent view of runway 14, the taxiways, and the airplanes from the tower cab. When the controller looked to the end of runway 14, the line of sight was directly over the intersection where the Cessna 172 was waiting. The controller would have had to look over the 172 to see the end of runway 14. Why didn't the controller see him? Did the controller, because of his belief that the 172 was on the other side, simply disregard anything he saw on the near side?

And the biggest question and the best possible prevention of the accident: Why did the Cessna 172 pilot pull out in front of an oncoming airplane on takeoff roll? The 172 pilot was the closest to the 152 and would have had an unobstructed, broad daylight, view back down the runway. But the pilot of the 172 also must have held a belief that the runway was clear just because the controller said it was clear. This tragedy can be so clearly broken down after the fact, but nobody saw the chain forming at the time.

The controller and the Cessna 172 pilot had momentarily lost their awareness of the situation. The controller had an aid to assist in keeping the big picture in the form of a flight strip. But the strip had not been completed properly as one controller departed the station and another took over. The Cessna 172 pilot believed blindly that the runway was clear just because the controller told him to taxi onto the runway. The Cessna 172 pilot was unaware of the other airplane's takeoff roll and apparently did not check for himself. Pedestrians, automobile drivers, and airplane pilots should always look both ways before crossing a street or entering a runway. Regaining awareness of the situation in this case could have been very easy, but it all happened fast and, as a result, four people died without leaving the ground.

It seems that the most obvious things can become elusive when you don't concentrate on the details. A controller and a pilot simply looked past the obvious and a tragedy took place. Looking past the obvious almost caused another tragedy in Charlotte, North Carolina, one night.

THE PILOT'S SIDE OF THE STORY: ASRS NUMBER 410622
Upon taxiing out from Charlotte, North Carolina, I picked up my IFR clearance from the tower and taxi instructions. The controller told me to taxi to runway 18L at intersection Alpha, which I did. The controller cleared me for takeoff from runway 18L at intersection Alpha and to fly a heading of 210 degrees, which I did. Later I was informed by the Charlotte approach control by telephone that runway 18L had been

closed, and that I came close to hitting a vehicle that was on the runway.

THE CONTROLLER'S SIDE OF THE STORY: ASRS NUMBER 410925   Airport operations closed runway 18L for maintenance. I turned off the runway lights and put the closure on the ATIS. Twenty minutes later I (working alone in the tower) taxied an aircraft to the closed runway and issued a takeoff clearance forgetting that the runway was closed. The aircraft departed over vehicles on the runway. Factors to include: (1) Airport operations—when closing a runway and having vehicles operate on the runway, should have better illuminated vehicles. (2) Pilot—the pilot did not listen to the ATIS or question why the runway lights were turned off. He also did not see the vehicle. (3) FAA—although the FAA knows runway incursions to be a problem they had no memory aids to alert the controller that the runway was closed.

This ground near-miss could have been a fatal accident for the pilot and the vehicle operator and there are plenty of questions that could be asked. Why would a pilot take off from a Class B airport like Charlotte at night without runway lights? That should have been very suspicious. And the controller seemed to blame a lack of vehicle lights, the pilot, and lack of memory aids instead of himself. As pilots we should remember this story whenever something does not seem right. Never be afraid to ask the controller about something you are not sure of. "Hey, are the runway lights on 18L inoperative tonight?" If the pilot had asked that obvious question, the threat would have been eliminated and everyone's awareness restored.

A similar situation once took place in the air, when a controller made an error, but the pilot did not ask the correct questions and a fatal accident was the result.

NTSB NUMBER LAX96FA078   The aircraft impacted mountainous terrain in controlled flight during hours of darkness and marginal VFR conditions. The flight was vectored for an instrument approach during the pilot's Part 135 instrument competency check flight. The flight was instructed by approach control to maintain VFR conditions, and was assigned a heading and altitude to fly that caused the aircraft to fly into another airspace sector below the minimum vectoring altitude (MVA). FAA Order 7110.65 (Air Traffic Control Handbook), Section 5-6-1, requires that when a VFR aircraft is assigned both a heading and altitude simultaneously, the altitude must be at or above the MVA. The controller did not issue a safety alert, and in an interview, said he was not concerned when the flight approached an area of higher minimum vectoring altitudes because the flight was VFR and "pilots fly VFR below the MVA every day." At the time of the accident, the controller was working six arrival sectors and experienced a surge of arriving aircraft. The approach control facility supervisor was monitoring the controller and did not detect and correct the vector below the MVA.

Probable Cause: The failure of the air traffic controller to comply with instructions contained in the Air Traffic Control Handbook, FAA Order 7110.65, which resulted in the flight being vectored at an altitude below the minimum vectoring altitude (MVA) and failure to issue a safety advisory. In addition, the controller's supervisor monitoring the controller's actions failed to detect and correct the vector below MVA. A factor in the accident was the flight crew's failure to maintain situational awareness of nearby terrain and failure to challenge the controller's instructions.

One of the easiest awareness traps is the overdependence on air traffic control. When first reading this accident report, it seems that this is clearly and solely a controller error accident. In fact, the controllers do bear the majority of this blame, but the pilots depended on the controller too much. When being radar vectored, pilots should keep up with their position. Don't just hold heading and altitude blindly. The pilots should have known that they were approaching higher terrain and asked the controller about it. The NTSB's probable cause states that a contributing factor of the accident was the "flight crew's failure to maintain situational awareness...." The pilots could have noticed that they were being vectored away and asked, "About how long will you need us on this heading?" Or an even more direct question, "What is the minimum safe altitude out here?" Either question would have helped the controller regain awareness before it was too late.

ASRS NUMBER 449887   While being vectored at 5000 feet, I was given a 340 degree heading from Burbank airport to an ILS approach at Van Nuys, California. I was in IMC but the controller neglected to issue any lost communications instructions nor did he inform us that radar contact had been lost with us over high terrain. We were approaching north of Umber where the minimum vectoring altitude (MVA) was 6000 feet. We turned west on our own and switched from frequency 120.4 to 134.2. The same fellow was working both frequencies. Radar contact was eventually restored and we landed at Van Nuys without incident. Someone was vectored into a mountain in the same area several years ago. I called Southern California radar control and talked to a quality assurance specialist who agreed that turning west and changing frequencies was the right action. We both agreed that a Remote Communications Outlet (RCO) or enhanced radar service in the area would considerably improve safety.

This pilot may have been referring to the previous accident report (LAX96FA078) when he said, "Someone was vectored into a mountain in the same area several years ago." The difference was that the second pilot maintained the awareness of his position and knew that he was headed for an area of higher terrain and MVAs. Awareness saved the pilot and any passengers in this situation.

Poor communications from controllers to pilots and from pilots to controllers can mask problems and create a situation where awareness is lost.

NTSB NUMBER NYC88LA199   The flight departed Wilmington, Delaware, on an IFR flight plan en route to Block Island. According to the pilot, as he passed Hampton VOR he saw fog obscuring the ground. Upon reaching the destination, Boston Center cleared the aircraft for the approach to Block Island. The pilot stated that he was unable to shoot the approach due to low fuel and requested assistance to a clear area or to an ILS approach. Boston Center vectored the aircraft to the Groton Airport where the pilot made at least four attempts to locate the airport. The aircraft was then vectored West Easterly, where once again the airport could not be located due to weather. The aircraft was being vectored to Providence when fuel exhaustion was experienced and the aircraft was ditched in the Narraganset Bay.

Probable Cause: The failure of the Boston Center controller to adequately understand the pilot in command's request and to assess the gravity of the situation.

The pilot was unhurt after the airplane struck the water and his passenger had only a minor injury, so this could have turned out much worse. But it remains a case where a pilot and a controller were talking but not communicating. The NTSB again in its statement of probable cause points to the controller's inability to "…assess the gravity of the situation."

There are times and places when the loss of awareness is predictable. One of these is when a controller handoff is made. A "handoff," as the name implies, is when the responsibility and control of an airplane transfers from one controller to another. This can happen either on the ground or in the air. Controllers have jurisdiction over certain geographic areas. When an aircraft flies to the edge of

one of these areas, the controller arranges for the controller of the next area over to "accept the handoff." When
the handoff is accomplished, the new controller takes
over, but the new controller does not have a "history" with
the new aircraft. The new controller does not know what
is going on with the flight just yet. The controller can catch
up by reading the aircraft's flight strip or asking questions
of the previous controller, but in the meantime there is a
built-in gap in awareness. Once a midair collision took
place that was literally in the gap between controllers.

NTSB NUMBER LAX93FA055A   After departure from
Sacramento, the pilot of a Cessna 210, had declined further
radar traffic advisories, and was climbing on about a
152-degree heading. The pilot had changed radio frequencies, and was receiving weather advisories for his destination airport. A Cessna 180, was in cruise flight at 3,100 feet
on about a 307-degree heading, and the pilot had just terminated his participation in receiving radar traffic advisories
from Stockton approach control. The two airplanes collided
near head-on about halfway between two radar facilities.
The right wing of the Cessna 210 contacted the left wing of
the Cessna 180, and both wings separated about midspan in
about a 90-degree orientation from each other. Visibility was
reported as 10 miles in haze. Other pilots reported that
smoke limited visibility from 1 to 5 miles in the area. Both
airplanes had Mode C altitude encoding, and both were
depicted on radar prior to the collision.

Probable Cause: The failure of the pilots of both airplanes to
see and avoid each other. Contributing to the accident was
the failure of the Sacramento controller to advise the
Stockton controller of the impending unsafe situation, and
the failure of the Stockton controller to issue a traffic advisory and a safety alert to the pilot of the Cessna 180.

Both pilots had voluntarily given away an opportunity
to remain aware when they declined radar assistance.
The controllers were faulted for not doing more to see

the midair collision coming and to warn the pilots, but one pilot had already vacated the frequency and took himself out of the awareness loop.

Then there is the famous case of the US Airways and the Skywest collision. A lapse of awareness and the inability to recognize the loss created the situation where one airliner landed on top of the other in Los Angeles, California.

NTSB NUMBER DCA91MA018A Skywest flight 5569 had been cleared to runway 24L, at intersection 45, to "position and hold." The local controller, because of her preoccupation with another airplane, forgot she had placed Skywest 5569 on the runway and subsequently cleared US Air flight 1493, a Boeing 737, for landing on the same runway. After the collision the two airplanes slid off the runway and into an unoccupied fire station. The tower operating procedures did not require flight progress strips to be processed through the local ground control position. Because this strip was not present, the local controller misidentified an airplane and issued a landing clearance.

Probable Cause: The failure of the Los Angeles air traffic facility management to implement procedures that provided redundancy comparable to the requirements contained in the National Operational Facility Standards and the failure of the FAA air traffic service to provide adequate policy direction and oversight to its air traffic control facility managers. These failures created an environment in the Los Angeles air traffic control tower that ultimately led to the failure of the local controller to maintain an awareness of the traffic situation, culminating in the inappropriate clearances and subsequent collision of the US Air and Skywest aircraft. Contributing to the cause of the accident was the failure of the FAA to provide effective quality assurance of the ATC system.

The US Air and Skywest collision was created by a controller who became "preoccupied," and this caused the "failure of the local controller to maintain an awareness of the traffic situation." The NTSB report also found that the environment that surrounded that controller actually assisted in the loss of awareness. On another occasion a similar situation occurred where one aircraft almost landed on another. ASRS number 421891 is the story from an anonymous air traffic controller.

I had just relieved the previous controller about one minute earlier and was trying to establish vertical separation between a previous aircraft departure and a VFR Cessna overflight [of the airport] before switching to aircraft over to departure [control]. I was also trying to turn up the radar scope's map and alpha numeric brightness which was very low when aircraft X called ready to go. I scanned the runway, but apparently not far enough on final. I cleared aircraft X for takeoff and resumed adjusting the scope when a short time later aircraft Y, a Lear on final approach who had been cleared to land by the previous controller, asked what aircraft X was doing. I looked up and saw the Lear (aircraft Y) on short final with aircraft X past the hold line with his nose over the runway edge. I told the Lear to go around. The Lear did not respond and continued to land. The Lear pilot apparently thought that aircraft X was at fault and said he caused a runway incursion. It was my fault. Had I set up the scope prior to taking the position, or simply told aircraft X to hold short while I was busy with the departing traffic and the VFR traffic, and taken an extra moment to fully settle into the position, this incident would not have occurred. I've heard that the first few minutes and the last few minutes on a radar position are when the most incidents happen in ATC, and now I know from experience. Be extra vigilant when taking the position and don't let the fact that there is very little traffic lull you into letting your guard down.

The pilot of the Lear Jet, described as aircraft Y, saw the other airplane as it entered the same runway he was landing on. But this took place during daylight hours. The US Air pilots in Los Angeles did not have that advantage. This was a close call, one that points out that pilots must maintain vigilance and be ready to overcome controller mistakes.

Recognizing the Loss of Situational Awareness

In every example cited in this chapter, accidents and incidents could have been avoided if a pilot or a controller had realized sooner what was taking place. The time can be short between the loss of awareness and disaster. Pilots must wake up and see the reality of the situation before they lose the ability to control the situation. Pilots need a set of guidelines to use during their flights that will point out when situational awareness is slipping away. The following are 12 keys that can help pilots recognize dangerous situations before all hell breaks loose.

The keys to recognizing the loss of situational awareness

1. Measure twice, and cut once.
2. Use both camera lenses.
3. Keep your thoughts in the airplane.
4. When you feel rushed—slow down.
5. Confirm incoming information.
6. Watch out after handoffs.
7. Silence is a trap.
8. Have a troubleshooting plan.
9. Have a plan for distractions.
10. When you are uncomfortable, it is for good reason.
11. Situations don't always follow the same pattern.
12. Visualize the "perfect flight."

1. Measure twice, and cut once.

This phrase is used in woodworking. If you have an expensive piece of wood and you want to make it into a table, it doesn't pay to be sloppy. If you cut the wood without an accurate measurement, you will most likely make a mistake and the wood will not fit properly. If you measure quickly and cut without double-checking, you will inevitably be buying more wood. It would be much better to measure the wood twice and make sure it is correct the first time. Measuring twice means you only have to cut once. This analogy holds true for flight planning and preparation as well. Situational awareness starts on the ground. If you do a sloppy job of preparation and jump into the air without accurate planning, your chances of becoming unaware of your surroundings are doubled.

Measuring twice to the pilot means preparing a good plan and then working that plan. You must get accurate and up-to-the-minute weather information for any flight. For flights away from an airport, even more work is needed. You must understand how wind will affect the flight, how long it will take to make the trip, and how much fuel is required. You must perform a thorough and suspicious preflight inspection. It is easy under normal circumstances to lose sight of the big picture, so when pilots take off without even the basic preparations completed, their odds of encountering problems magnify. The concept of measuring twice and cutting once is just insurance against making in-flight situations worse.

2. Use both camera lenses.

In Chapter 2 the example of a camera was used to show how attention should be focused. While in flight you should constantly be shifting your attention for the big picture to small details and back. Never concentrate on any one item for very long. The danger is that while you hold your attention on one item, job, instrument, or task, something else is going awry without your seeing it. Something could go wrong when your back is turned. Therefore you can never turn your back on anything for very long. Pilots can detect problems and maintain awareness best by focusing on the correct item at the correct time, but also by being able to revolve their focus. By figuratively changing lenses from wide angle to zoom, the pilot is constantly paying close attention to one detail, but doing so while seeing everything else.

It has also been compared to juggling. When people are juggling three, four, even five objects at a time, they must catch and throw the individual items (the details), but they must also keep up with the timing and coordination of every item (the big picture). If the juggler fails to constantly shift focus from details to the wider view and back, a mistake will be made and the objects will all hit the floor. When you are flying an airplane, in choppy weather, into a busy airport, with rapid controller conversation, and a sky filled with other aircraft—you are juggling. In order to keep things from falling apart, you must keep your attention moving in and out—from wide angle when looking for traffic, to zoom lens when changing radio frequencies, back to the big picture when it comes to understanding where you fit into the approach sequence, to the detail of setting the engine power for a descent.

If at any point you get the feeling that things are getting away from you—that you are losing awareness—you must increase the speed in which you switch the lens. Of course, it is very possible to lose awareness and not know that you have lost anything. To be sure, pick up the pace of your focus when the workload picks up.

This will give you the best chance to keep up and recognize the loss of awareness.

3. Keep your thoughts in the airplane.

When I drive my car, my mind can be a million miles away. I have often arrived home and could not exactly remember making all the correct turns and stopping at all the lights. It is possible to do the same in an airplane, but it is just not safe (probably not safe in a car, either!). Pilots must learn to compartmentalize. This means you must learn to shut out all thoughts that do not pertain to the flight problems at hand. You must place all other life-problems in a box and leave them closed up until the airplane is again tied down. Pilots who do this are the same pilots who describe flying as an "escape." It should be an escape from all other worldly cares. You simply cannot fly safely if you are thinking of paying the bills, or relationship problems, or work conflicts.

Airline crews fly using a "sterile cockpit." This means that below 10,000 feet, all conversation must be devoted to the aviation tasks at hand. At low altitude those tasks would be takeoff, departure, arrival, instrument approach, and landing. These items are the most pilot-labor intensive. Prohibiting conversation about nonrelated items during these times reduces distraction and intensifies focus. Every pilot should practice a form of the sterile cockpit concept. Inform passengers of when pilot workload will be high and ask them to keep conversations to a minimum during those times. But even when you fly alone, you should keep attention directed to the airplane and the situation.

Any flight can be divided up into at least three phases: takeoff and departure, en route, and airport arrival. Divide your thoughts up the same way. Don't leave the "sterile cockpit" until all the duties of one phase have been completed. This will help you stay focused and plan ahead to the next phase. There are plenty of outside distractions that will threaten your awareness, so don't compound the problem by adding self-imposed distractions.

4. When you feel rushed—slow down.

There are times when fast action is required, but there is a difference between being rushed and being quick. Being rushed is a state of mind. The feeling is often associated with a flood of duties or responsibilities. You can have so much to do at once that you do not know where to begin. You get the sensation of being swept away, unable to keep up, and out of control. The rushed feeling can quickly turn into a panic feeling if the individual items are not addressed. When you feel rushed, you must mentally slow down and attack the problems one by one. It would be best to prioritize the problems so that you attack the most pressing item first. Eventually you will start to regain control of the situation and ultimately solve all the problems.

An instrument approach offers many challenges for pilots, especially when flying alone. Among the duties that must be accomplished are radio communications, chart selection and reading, radio frequency tuning and identifying, aircraft control (which can include power, speed, and altitude changes), and the ability to know the aircraft's position relative to the approach at all times. All of these duties must be accomplished in a very short time, so the rushed feeling is common during instrument work. Unless you slow down to the point where each and every item is processed, the approach will fall apart. Remember, the phrase "slow down" as used in this context is a mental process where brief concentration time is given on a priority basis. You cannot take all day to accomplish these tasks; the actual elapsed time may be very short, but the systematic approach to the problem relieves the rushed feeling. If you ever get the chance to observe professional pilots at work, you will notice that they never seem rushed, but they are always busy. This is because they prioritize the duties at hand, solve the problems that come up one by one, and plan ahead.

5. Confirm incoming information.

Always make decisions based on more than one source of information. Just like a reporter seeking confirmation of a story before it is printed, you should always look for confirmation before you take action. The Cessna pilot in the Sarasota-Bradenton accident heard the controller tell him to taxi onto the runway, but did not confirm this by looking both ways. The pilot relied on only one source of information—the controller's instructions. If he had looked for traffic himself, and in doing so confirmed the correctness of the controller's instruction, he would have seen the mistake and an accident would have been prevented. In almost every accident example, the tragedy could have been avoided if a pilot or a controller had one more piece of information. Knowing this in advance, pilots should never act on a single bit of information. Look to corroborate all incoming information.

Once a pilot saw that his oil pressure gauge was suddenly reading zero, which could mean the engine had no lubrication, the pilot immediately shut down the engine and headed for a field. The airplane was damaged during the landing, and in the postaccident inspection the oil pressure gauge was found to be inoperative. The engine had been receiving the proper lubrication the entire time. The pilot acted on only one bit of information, and as a result had acted inappropriately. If the engine was indeed losing oil, confirmation would soon be seen on the oil temperature gauge. As oil is lost, engine friction goes up and overheating would be seen, which would confirm the problem. If the pilot had made a cross-check or confirmation, he would not have risked an off-airport landing and damage to the airplane.

Navigation situations require confirmation. Many pilots have gotten lost because they did not double-check their position. When you are looking for a check-point, like a highway, it is easy to believe that the highway you see is in fact the highway you are looking for. But the moment you trust false information is the moment your awareness escapes. To regain awareness, you can't jump at the first thing that comes along. You must verify. Is the highway you see aligned properly? The point where you should cross the proper highway is on a particular VOR radial. Is this highway on that radial? These are the questions you can ask to seek con-firmation. If you abandon your original plan and strike off in a new direction following an unconfirmed high-way, you will soon be off course, and unaware.

Confirmation can come from other instruments, other navigation aids, or even a call to ATC. In a way it's another form of "measure twice, then cut once." If you make an accurate assessment that is backed up by con-firming information, the actions that you take will be more appropriate. The worst form of unawareness is when you don't realize that you are unaware. For this reason, pilots should seek confirmation of decisions that they make at all times. Maybe by seeking confirmation you will uncover a mistake that had gone undetected. Making the unknown become known is the same as regaining situational awareness.

6. Watch out after handoffs.
A handoff is controller-talk for when an aircraft passes from the control of one controller to the next. In radar

rooms across the country, controllers sit in darkened rooms facing their radar screens. In most facilities the screen they are watching is actually a computer screen that displays a portion of the country. It may be that controllers sitting side by side are controlling adjacent airspace sectors, or it may be that the next sector is controlled by a person who is sitting at a screen several hundred of miles away at another ATC facility. When an aircraft flies off one screen and onto another screen, they can only do so if the new controller accepts a "handoff" from the previous controller. Like a video game, the previous controllers move a cursor with a track ball over the airplane that is leaving their airspace. They signal the computer to make the airplane's symbol start to flash. There is a small overlap between radar screens, so the new controllers will see a flashing target appear on the edge of their screen. To accept the handoff, they must run their track ball up to the flashing target, make a keystroke, and the handoff is accomplished. At the same time, information about the new aircraft is relayed on a strip of paper. On the newest computer/radar installations, the flight information appears on a separate computer monitor. The information about the aircraft is passed from one controller to another like e-mail. From the paper strip or computer monitor, the new controllers get information about their new customer. For the new controllers to "get the picture," information must change hands. Sometimes this exchange of information is delayed and this causes a gap in awareness on the controllers' part immediately following the handoff. Have you ever received a handoff while flying an assigned heading and the new controller asked about that heading? "1234A say your heading" the controller might ask. You say, "Heading 270 assigned." When you add that last word, "assigned" you are telling the new controller that the previous controller had placed you on that heading. The fact that the new controller does not know about the heading assignment is an example of a momentary breakdown of the information flow following a handoff.

Pilots should be aware of this as a potential problem. When you fly for several minutes across the screen of one controller, you start to get to know each other. You make requests of the controller to change altitudes, or to deviate around weather, to change a route of flight, or even change your destination. You and that controller develop a history together. When you are handed off to another controller, the new controller doesn't know all that has happened in the recent past, so that the controller cannot be completely aware of your situation. It's like having a good next-door neighbor. Over time you get to know about your neighbor and you have helped each other out from time to time. Then one day your neighbor moves and a new family moves in. It's awkward at first until you get to know the new people. During the transition with new neighbors and new controllers, there can be miscommunications and problems.

When possible, try to predict a handoff. If you know that a handoff is imminent, you should wait to make requests with the next controller. If you were interested in an altitude change, but asked for the change just prior to a handoff, the controller would probably tell you to make that request with the next controller. But if the altitude change was granted and then a handoff was made, you would be climbing or descending into a new controller's area. If this altitude change information had not been passed along, then the new controller might not be aware of what is going on. It would have been better to have waited and made the original request with the new controller. In this way, the new controller is in on the altitude change from the beginning and maintains awareness.

How can you anticipate a handoff? The area around Class C, Class B, and the few remaining TRSA airports are usually divided in half. The dividing line is usually the alignment of the major runways. Picture a circle with the airport in the middle. At the airport the major runways run north and south: runway 36L and 36R/18R and 18L for instance. The north-south alignment of the runways would divide the circle in half, forming an east sector and a west sector. As you fly through the area, use the major runways to picture this dividing line. You can expect a handoff as you near that line. The different sectors will have different approach control frequencies. These frequencies are shown on a sectional chart in an approach control box. There will be a different box for each sector, and they are usually printed on opposite sides of the airspace surrounding a major airport.

On an IFR en route chart, the boundaries between the different Air Route Traffic Control Centers (ARTCC) are depicted. When approaching one of these boundaries, it is easy to predict a handoff to the next center. The ARTCCs are made up of about 20 smaller sectors, each with a different controller and frequency. These sectors and frequencies are also on IFR en route charts, but their exact boundaries are not shown. Even without the boundary printed on the chart, you can anticipate a handoff as you get closer to where a new sector's frequency is shown on the chart. All of these are ways to expect the handoff and be ready to heighten your own awareness when the controller's awareness has the potential to diminish.

Pilots can help the controller make a smooth transition, and therefore maximize awareness, with good communications. Be ready to tell the new controller in your first communication what you are doing. By doing this you are helping to fill in the awareness gap that may exist as the new controller brings you on. Your first transmission might be, "Atlanta Center, this 1234A out of 2500 for 8000." This gives the new controllers immediate information that they may already have, but they also may not have. If there is crossing traffic in this new sector at 4000, I just told the controllers of a possible conflict that they are now aware of. Another example takes place when approaching an airport and being handed off from approach control to the control tower. "Memphis tower, 1234A on a visual approach to runway 27 and I have the traffic to follow in sight." This tells the tower controller on the first transmission that you have the airport in sight, you have the traffic in sight, and you already know which runway you are landing on. There is also a gap to fill after takeoff from a controlled airport. Soon after liftoff, the tower controller will say, "1234A contact departure." When they say that, it means that your control is being passed from up high in the control tower where the controllers can actually see you with their eyes, to the radar room downstairs where the controllers see you only on radar. To the radar controller, you have come from nowhere. You have just appeared on the screen, so you must help them gain awareness of what it is that you are doing. "Richmond departure, 1234A, off runway 34, from 1500, in a left turn to 240." This gives the controller information about where you are and where you are going instantly. All of these are examples of how pilots and controllers help each other out by maintaining awareness.

Whenever you get a controller handoff, you must raise your level of awareness. Many accidents have taken place when an aircraft's control was passed and the receiving controller had not yet built the big picture. Help the controller out with good initial communications and keep your eyes wide open.

7. Silence is a trap.

Even on an average traffic-load day, an air traffic control frequency will have a continuous conversation going on. There will very seldom be long breaks in the transmissions going back and forth between pilots and controllers. So if you ever notice a long period of silence on the frequency, it should be a warning sign. We have all done it. We have turned down the volume of the radio being used for ATC communications so that we could listen to AWOS or identify a navigation frequency. Then we forgot to turn the volume back up. For several minutes controllers call, but we don't hear anything. An unusual period of silence on the radio can mean that things are taking place around you that you are unaware of. You did not mean for it to happen, but suddenly you have completely lost situational awareness and don't even know it yet. The controllers could be calling you about conflicting traffic, or altitude or heading changes. Since you did not hear, you may have missed a controller handoff and now you are flying through someone else's airspace without knowing it. Check the radio frequently to ensure that the volume is up when radio congestion is light. If you haven't heard a controller for a while, you can ask for a radio check or make up a question for the controller: "Washington Center, this is 1234A. Do you have a ground speed readout on us?" You would not want to ask a low-priority question like that when the radio congestion was high, but you could ask when there has been a period of silence just to confirm that everything is working and that you have not missed anything.

In addition to having the volume too low, radio silence can be the result of improperly hooking up headsets, or the improper placement of the toggle switches on an audio panel. Radio silence may be due to using an improper frequency. I entered the traffic pattern at an uncontrolled airport recently and reported my downwind, base, and final leg. I landed, but I had heard nobody else on the unicom frequency. Once inside the FBO, I saw that the desk attendant had a unicom radio, so I asked if he had heard me on my way in. He said he had not, but that they had recently changed to a new unicom frequency. They were using 123.0, and I had used 122.8 from off the sectional chart. I did not know of any conflicts but I had approached the airport and flown the complete traffic pattern with an uncomfortable radio silence. The next edition of the sectional chart had 123.0 listed.

To prevent radio frequency mismatches, use the Common Traffic Advisory Frequency (CTAF) at every airport. The CTAF is indicated on sectional charts with the letter C on a blue ball. Next to the ball will be the frequency to use anytime, day or night. Sometimes the CTAF is a tower frequency used by a part-time control tower. When the tower is open you use the frequency and say, "Champaign Tower." But later when the tower is closed the same frequency is used and you should say, "Champaign Traffic."

Sometimes late at night when the traffic is very light, the amount of conversation can be minimal, with long breaks between talk. But don't let this lull you to sleep literally or figuratively. Check in with the controller periodically to ensure that the silence is not a mistake. Usually at 3 o'clock in the morning, the controller will not mind the extra conversation.

8. Have a troubleshooting plan.

Pilots need to do a better job of understanding the operating principles of their airplane's systems. I have known some pilots who have taken the attitude that airplane systems were just not their problem. After all, they were not A&Ps. Well I'm not an A&P either, but the safety of my flight depends on the detection of system problems that might arise. If a system or component does malfunction in flight, the pilot must:

1. Detect the problem.

2. Determine if the problem threatens the safety of the flight.

3. Take appropriate remedial action—if there is any to take.

4. Decide to continue the flight, land as soon as practical, or make an emergency landing if the situation warrants such drastic action.

Pilots who are not familiar with their airplane systems may not be able to complete even step one, which is to detect the problem. A private pilot with an instrument rating once took off into low-IFR conditions on a night flight of approximately 200 miles. At some time during the flight, the airplane's alternator went offline and electric power started being drained from the battery. The pilot did not detect the problem until the panel lights started getting dim and the controllers were unable to hear his transmissions. The pilot had lost situational awareness because he did not do a good job of systems detection. Had the pilot noticed the problem earlier, he could have recycled the master switch, which would have reset the alternator. It was discovered later that the alternator was working properly but had been taken off line due to a momentary power surge in the system. The story ended happily due to pure luck. The airplane temporarily flew out of the clouds over an airport. The pilot landed safely, but by this time had no electricity to operate the landing light or the flaps.

Pilots who do not know their aircraft systems well are really reducing their chances of a happy outcome when a problem occurs. The solution to this problem is to study the aircraft manual and ask questions of your flight instructor. The best idea is to go spend some time with the A&P technician who maintains the airplane. You will learn more about systems in 60 minutes with the A&P than you will reading any book.

Anytime a system or component fails to function as it is intended, the pilot must become aware of that fact as soon as possible. Then the pilot must have enough knowledge or familiarity with the system to diagnose the problem and take any action necessary to reduce the danger. It is a terrible feeling to be in flight with a problem and not know what to do about it. A pilot who cannot detect, diagnose, and reduce a systems threat is a pilot who has no defense against the loss of situational awareness. Being able to detect a problem, on the other hand, is the first step in regaining situational awareness.

As a flight progresses, take the time to make occasional system health checks. I try to make one of these checks at least after crossing every check point along the way. Look for normal operation of the engine by checking oil pressure and temperature gauges. Listen for changes in engine sound and vibration. Use carburetor heat occasionally or continuously when conditions warrant. Check the electrical system by observing the ammeter or loadmeters. Verify that the flight instrument systems are functioning properly by checking the vacuum gauge and deciding on the need for pitot heat. Make sure you update your altimeter setting and match the magnetic compass with the DG as your flight continues. Verify that mixtures have been set properly after climbs and descents. All of these checks are just preventive maintenance that guards against losing situational awareness of the airplane itself. Just as with humans, the earliest detection of a health problem produces the best chance for a cure. Do not fly again unless you are confident that you can detect and troubleshoot your systems.

If you ever detect a systems problem, remember that you must maintain aircraft control at all times as you attack the problem. Being distracted by a problem has led to many control-loss and stall-spin accidents. Equally divide your attention between flying the airplane and thinking through the problem. Be ready to

1. Detect.

2. Determine the level of threat.

3. Take appropriate action.

4. Make plans to get on the ground.

9. Have a plan for distractions.

It is not a matter of if, but of when. At some point you will be faced with a distraction in flight. You should not be caught by surprise by a distraction, but instead shift into a preplanned course of action to deal with the distraction. Your number-one priority is always to fly the airplane. This means that airspeed must be maintained throughout the challenge of any distraction. You cannot allow a distraction to turn your full attention in only one direction. Instead, when a distraction takes place your attention must be like a ping-pong ball bouncing quickly back and forth between the airplane and the distraction.

Once the forward baggage door popped open and a suitcase flew out from a light twin-engine airplane on takeoff. The suitcase hit the left engine and the engine lost power. It was the classic multiengine takeoff situation. Students in multiengine airplanes practice this engine-out scenario many times: throttles forward, dead-foot/dead-engine, reduce drag, bank slightly into the good engine, maintain the blue line, and climb out to safety on the good engine. But in this case the pilot turned his full attention to the distraction of the suitcase hitting the left propeller and never completed the engine-out procedure. Quickly the airspeed dissipated below the minimum control speed and a fatal accident occurred. The pilot became surprised by the distraction and this robbed him of the ability to take action. The accident could have been avoided if the pilot had been able to overcome the initial surprise and go to work on the problem. You cannot help but be surprised some-times, but what happens in the next second will make the difference. Don't panic; just go to work.

After the initial surprise has taken place, the distrac-tion must then be prioritized. Is it something that can wait until after landing? Is it something that will force you to land sooner than was originally planned? Is it something that will threaten the immediate safety of the flight? Answer these questions, and the level of the chal-lenge will become clear. The problem may be evident or it may be hidden. If you are not sure what the prob-lem is, then while moving your attention back and forth between the airplane and the distraction, determine what is not broken or not wrong. Systematically cross off all the items that are not causing the problem.

A distraction may not threaten the safety of the flight and may simply be a momentary diversion of attention. This is like the juggler. The juggler is doing fine while juggling four objects, but then unexpectedly somebody throws in a fifth item. The juggler must blend in the fifth object but cannot lose concentration on the original four. When a momentary distraction occurs you must "take it in stride."

Pilots are tested for their ability to handle distractions. Every FAA Practical Test book has a paragraph in the opening chapter about the use of distractions during the flight test. The pilot examiners are supposed to deliberately create distractions during the flight test so that they can evaluate how the pilot-applicant handles the distraction. The distractions presented are supposed to be realistic but reasonable. Remember, with distractions it's not if, but when, and this is especially true on the checkered.

Have an advance plan to deal with the distraction:

1. Overcome the initial surprise.
2. Maintain aircraft control and take the distraction in stride.
3. Prioritize the distraction.
4. Take appropriate action.

10. When you are uncomfortable, it is for good reason.

Most pilots have had better than adequate training. Most pilots have good common sense, and most pilots have good instincts. Training, common sense, and instinct join together to create a comfort level in the consciousness of the pilot. Comfort level is not a scientific expression that can be quantified. It is more a "gut feeling." You may have said, "I don't have a good feeling about this." That "feeling" is not always something that you can put your finger on, but you know something is just not quite right. Whenever you have that feeling—trust it!

That not-quite-right feeling may be your first warning that something, as yet hidden, is taking place. You do not have hard evidence that points to anything specific, but you are just not comfortable. When this feeling hits, start looking around. Look at engines and systems, look at navigation frequencies and settings, look at instruments, and verify radios.

Sometimes that uncomfortable feeling comes from expectations. Maybe the controller has asked you to "keep your speed up" in order to fit into the flow of traffic at a time when you normally would be slowing down. Maybe you hear another airplane report on downwind when you are already on the downwind leg.

Sometimes that uncomfortable feeling can come from being rushed either in the air or on the ground. Have you ever been in the middle of your pretakeoff checklist and looked back to see another airplane waiting behind you. Other pilots have ways of letting you know that they are ready to go and that your careful checklist is holding them up. They sometimes taxi up real close, or rev the engines several times, or even get on the radio and say, "How much longer are you going to be?" This pressure to hurry along can produce that uncomfortable feeling. Trust that feeling—don't get pressured into the air before you are ready.

Sometimes that uncomfortable feeling can be produced by another pilot that you are flying with. It is possible for two pilots in the same airplane to have a difference of opinion about a procedure. If you are ever with another pilot and you are not comfortable about how things are progressing, don't be afraid to say, "You know, I'm just not comfortable with...." Saying it that way will not challenge the other pilots but might open their eyes to a problem that you see. This scenario plays out every day among flightcrews. Remember the example of the flightcrew that landed in Brussels instead of Frankfurt. At several points along the timeline of that incident, one or more members of that crew became uncomfortable with one item or another. But at no time did any of them put their foot down and say, "You know, I'm just not comfortable with these fuel amounts." Nobody spoke up to say, "Hey, I'm not happy that we are using a VOR that is not on our chart." Any expression like that could have snapped the crew back to reality—but nobody trusted their uneasy comfort level.

If you think that something is not quite right—the chances are good that something is in fact not right. Don't ignore your gut feelings. Trust your common sense and instincts. In the aircraft we have warning lights and warning horns that are supposed to give us an early heads-up to a problem. Your comfort level should also be considered a built-in warning system for problems.

11. Situations don't always follow the same pattern.

No matter what we say to others, we do like patterns, routines, and habits. Patterns, routines, and habits produce situations that we are familiar with and therefore comfortable with. As pilots we really don't want to face the unexpected. We want everything about a flight to go as predicted. If you ask another pilot, "How was your trip?" and the pilot answers, "It was uneventful," then you know everything went as expected and it was a good trip. But sometimes our reliance on expectations can get us into trouble. Eventually there will come the flight when things don't go by the book. On that flight what we expect to happen may shield us from seeing what is actually taking place. If you expect fellow pilots to enter the traffic pattern on a 45-degree angle to the downwind leg, then your awareness to a pilot coming in on the base leg will be reduced. You don't expect another airplane to be coming in on base, so you don't ever look in that direction for traffic.

You become familiar with the routine in and around your home airport. From repetition you learn the frequencies, the taxi procedures, the approaches. Being familiar with the situation is good, but as always there is a line that can be crossed. It is possible to get too comfortable. Sometimes when airports have parallel runways, over time one of those runways unofficially becomes the general-aviation runway and one becomes the air-carrier runway. Usually that is because of the runway lengths or proximity to the terminal or parking. You become extremely familiar with taking off and landing on one of the two runways. Then one day you are assigned the other runway, but your expectations and past routines are powerful, and it doesn't sink in that this time will be different. This is how pilots find themselves at the wrong place at the wrong time.

To maintain awareness or regain lost awareness, we must expect the unexpected. Every day is a new day, with different challenges than the day before. Expectations, routines, and habits that are relied upon too heavily can create a dangerous overcomfort. The next time you fly, try to discover at least three things about that flight that had never happened before. By looking for the differences, you will maintain an awareness that otherwise would have been masked by other expectations.

12. Visualize the "perfect flight."

To maintain awareness before and during a flight, time must be taken. Time should be taken before getting started to visualize the flight from start to finish. No, you cannot predict every radio transmission, every turn, or every needed deviation, but you can remember what you did on past flights that made them go smoothly. Picture the route of flight and the checkpoints that you will pass. Place any weather threats into your picture and ask yourself, "what if...." What if I cannot go directly to the destination? Which way would be the best direction to divert? What other airports lie along the route that I could use if something unexpected came up? What charts will I need? What airspace types will I penetrate? What controllers will I be required to communicate with? What controllers could I voluntarily communicate with? How many ways can I find along the route to communicate with the FSS? There could be another hundred such questions to ask.

After you have taken the time to mentally fly the perfect trip, it will become easy to detect differences between the perfect trip and the actual trip. Anything that actually takes place that did not take place on your "virtual" trip should be a warning sign. It has been said that it is hard to see a problem that is not there. This means that a problem can exist under your nose, but if you don't know to look you will not find it. But if you have a model to go by—the imagined perfect flight—any differences that come up will be keys to better awareness.

If the controller assigns you to a heading that takes you in a direction that you did not expect—a difference between the actual and perfect flight has occurred. This difference should signal you to the fact that something extra is happening that you need to be aware of. Maybe this is a vector for traffic that you should be looking for?

You don't expect the engine rpm to begin fluctuating. Engine problems would not have been part of the perfect scenario. So any difference in engine or system performance should be your signal to shift into an awareness high-gear. Don't ignore problems in hopes that they will just go away—attack the differences.

Visualizing an instrument approach is one of the best ways to plan for the approach. Instrument approaches are filled with heading changes, power changes, altitude changes, timed segments, wind correction, and the need for planning. Seeing the approach in advance and comparing it to what actually is happening is the fastest way to detect problems and make adjustments.

When the flight has been concluded, go back over the trip in your mind and see where differences took place and what you did about them. Reviewing a past flight will help you visualize the next flight because you will have new experiences and circumstances to factor in. By seeing the flight in advance and using your past experience, you will gain the elusive quality of seasoning. Seasoned pilots rarely get caught off guard. They see things coming and are already working on the problem in advance. Seasoned pilots rarely lose awareness, but they have the ability to quickly regain awareness if it ever slips.

5

Recovering Situational Awareness

Much has been discussed so far about being "ahead" of the airplane, but pilots also know what it feels like to be "behind" the airplane. The first time you fly an airplane that is a little more advanced than what you are used to, you get the "behind" feeling. I remember my first take-off in a complex airplane. Between the landing gear retraction, faster speed, manifold pressure adjustments, and propeller control settings, I never looked outside the airplane once. I thought, "I'm glad this flight instructor is in here taking care of everything." I was so far behind the airplane, it felt like I was sitting in the back seat. Getting that "behind the airplane" feeling is a sure sign that situational awareness is gone. So when you realize that you have lost it, how do you get it back?

Accidents are often referred to as a chain of events, or a poor decision chain. During the sequence that leads up to an accident, poor decisions are being made because pilots are not acting with all the facts. They don't have all the facts because they are not completely aware of what is taking place around them. Regaining situational awareness can be accomplished by breaking the chain.

In every accident or incident sequence, there comes a point where the pilot's view of the situation and the reality of the situation separate. What the pilot knows and what there is to know become different. The pilot's perception of the situation and reality divert away from each other. This moment is called the "point of diversion" (POD). Downstream from the point of diversion, an accident or unfavorable outcome is bound to happen. The only way to recuperate is to rejoin perception and reality—to wake up or regain situational awareness.

There are times when the point of diversion is obvious; other times it is subtle. Sometimes the POD takes place immediately before the accident with little time to regain awareness. Other times the POD takes place and then a long time passes before the accident or event. Pilots can learn to regain awareness and redirect the destiny of the flight by identifying a point of diversion.

Like everything else, we can improve our ability to spot a POD with practice. To help practice, I selected the following ASRS and NTSB reports to be used as case studies. In each case, the pilot or pilots arrive at a point of diversion, but keep right on going without noticing it. Let's see if you can spot the PODs. I grouped the cases into broad categories so it will be easy to compare. The categories are Management of Critical Commodities, Weather Decisions, and Aircraft Operations.

Critical Commodities

ASRS NUMBER 441450 Upon landing at my destination, I discovered that the aircraft required 45.2 gallons to fill. The Pilot's Operating Handbook lists 48 gallons as the usable fuel with a minimum of 6.2 gallons per hour (GPH). The flight was originally planned to take 3.5 hours, burning 9.5 gallons per hour. Deviations to avoid weather stretched the actual time en route to 4.1 hours. Calculations en route suggested adequate fuel and the fuel gauges showed $\frac{1}{8}$ tank upon landing. I calculated fuel flows for the entire trip after it was over. I have concluded that the line person that filled the tanks prior to the above leg must not have filled them. The Cardinal looks like a Cessna 172 but has a thinner wing. I checked the fuel at preflight by dipping my finger in the tanks. It must have been an inch or so lower than normal, but I did not notice.

At least one POD took place before the engine was ever started. Never depend on anyone else to verify your fuel load. The pilot concluded that it was the lineman's fault for not completely filling the fuel tank. The lineman may not have completely filled the tanks, but checking the fuel by "dipping my finger" was inadequate. Discovering tanks with less fuel in them than planned before takeoff would have prevented this close call.

ASRS NUMBER 442960 I had fuel when I departed Ohio, due to a visual check. The gauges indicated, ½ tank in the left and ¾ in the right tank. I burned all the fuel in the left tank and then switched to the right. Upon crossing over Altoona Airport to enter the traffic pattern the engine started running rough. I switched tanks, turned on the pumps, increased the mixture, and informed unicom of the problem and requested to land. The engine continued to run rough. I landed, parked, and looked in the gas tank to discover that there was no fuel. The right gauge indicated ¼ tank when it was empty. I relied on my fuel indications and not on my judgment. I should have topped off and will before any flight again.

There really is only one reason why you should ever leave the ground with less than full fuel tanks—excess weight. Often a load of passengers, baggage, and fuel will exceed the aircraft's maximum gross takeoff weight. If you cannot leave any people or baggage behind, then the only thing left to reduce is the fuel. But anytime you calculate a fuel reduction for weight, you should at the same time plan on an extra fuel stop. If you can go with full tanks and still be at or under gross weight, then always top the tanks.

This pilot faced a POD when he elected to take off with "$\frac{1}{2}$ tank in the left and $\frac{3}{4}$ in the right tank" instead of full tanks. The pilot had another close call with fuel, but says he will always top off in the future.

NTSB REPORT NUMBER ATL89LA202   The airplane engine experienced a loss of power during final approach, and the airplane subsequently crashed into trees short of the runway. The student pilot was on a local, unsupervised solo flight and had concluded several touch and go landings prior to the power loss. The total flight time of this flight was about one hour. The operator reported that postcrash examination of the airplane revealed no fuel remaining in the fuel tanks. The operator noted that the airplane had been operated a total 4.6 hours since its last refueling. He also noted that the pilot who had flown the airplane on the previous flight reported that he had estimated the tanks were about $\frac{1}{2}$ full when he preflighted the airplane prior to this flight.

Probable Cause: The pilot's misjudgment of the airplane's fuel supply during preflight inspection, and his subsequent failure to monitor in-flight fuel consumption, which led to the engine power loss due to fuel exhaustion. Contributing to the accident were the pilot's decision not to refuel the airplane prior to the flight, and his lack of total flight experience.

This pilot did not have a close call, but an actual accident. The airplane was destroyed but the student pilot was unhurt. The student's point of diversion took place before takeoff when he thought there was enough fuel, when in fact there was not.

NTSB REPORT NUMBER ATL96LA091   The pilot reported that the fuel gauges were inaccurate, registering less fuel than was actually in the fuel tanks. According to his accident report, the pilot took off with five gallons of fuel in the airplane. He flew to a local airport and made a low pass. He then proceeded to another airfield and made a touch and go landing. The pilot then contacted approach control at his destination for final landing. According to the pilot, traffic was unusually heavy and the flight was vectored in sequence for landing. On final approach, the pilot advised the controller that the engine had quit, due to lack of fuel. Subsequently, the airplane collided with a powerline, then crashed to the ground in a residential area, about one mile north of the airport. During examination of the wreckage, approximately eight ounces of fuel were found in each fuel tank. The fuel caps were in place, and there was no evidence of fuel leakage from either tank.

Probable Cause: Improper planning/decision making by the pilot, which resulted in fuel exhaustion and loss of engine power, due to an inadequate supply of fuel.

This was another point of diversion that took place on the ground before the flight began. The pilot of this Grumman AA-1C suffered no injury in the crash, but jeopardized himself and others on the ground (residential area) because he did not recognize when his situational awareness was lost.

The pilot reports in this series have had one thing in common: The pilots were using the fuel gauges to determine quantity. This is a bad idea. Fuel remaining should always be calculated using elapsed time, not the fuel gauge indication. Fuel gauges are universally unreliable. In fact, the only time a fuel gauge is required by law to be accurate is when the tanks are empty. This fuel gauge law is hard to find. It was originally written into the Civil Air Regulations (CAR) in the 1950s. Unless a regulation is specifically superceded, it remains in force and that is the situation with CAR 3.672. The CAR (not FAR) rule 3.672 states that, "Fuel quantity gauges shall be calibrated to read zero during level flight when the quantity of fuel remaining in the tank is equal to the unusable fuel supply...." So any amount of fuel in the tanks may be misrepresented on the fuel indicator and still be legal to fly. But just being legal does not mean it is reliable and does not mean it is safe.

ASRS NUMBER 441635 The flight in question was the return leg of a cross-country flight from 3SQ to COQ. The aircraft suffered fuel starvation. An off-airport landing was made with no damage to the aircraft and no injury to the pilot or passenger. There was no threat to persons or property on the ground. Prior to departure from COQ, the pilot obtained a weather briefing from FSS by telephone and reviewed weather graphics from a private weather vendor. An IFR flight plan with an estimated time en route of four hours was filed with the FSS briefer. The time en route was based on a winds aloft forecast for the route which indicated that winds across the route were to be from the west-northwest with velocities from 10 to 20 knots. Based on the course direction for the flight (165 degrees magnetic on average), the forecasted winds should have resulted in a crosswind or slight tailwind. The aircraft had been topped off with 48 gallons of usable fuel after arriving at COQ from 3SQ and was verified full before takeoff. The True Airspeed was estimated to be 118 knots over a distance of 280 nautical miles, resulting in a no-wind time en route of four hours and four minutes. The flight planned fuel consumption was estimated at an average rate of 9 gallons per hour for the 4-hour flight requiring a total of 36 gallons. That should have left 12 gallons or approximately 1.25 hours in reserve. The groundspeed en route was less that anticipated based on the preflight information. About 4 hours into the flight and while descending through 3000 feet from a 5000 foot cruise altitude, the engine began to stumble. The fuel gauges appeared to indicate that between 5 and 10 gallons were remaining in each tank. The pilot switched tanks, turned on the auxiliary fuel pump, and the engine regained power. The pilot advised ATC of a fuel emergency and requested assistance. However, shortly thereafter, the engine began to stumble again and eventually stopped. The conditions were VMC at the time. The pilot established a best glide speed, picked a landing spot, looked for car/truck traffic and landed without incident on a hard surfaced farm road. The pilot then contacted FSS by telephone and advised that everything was OK. The pilot contacted personnel at the Jerseyville, Illinois airport and had 10 gallons of fuel delivered to the farm where the aircraft had landed. After refueling with the 10 gallons, local residents and Jerseyville airport personnel blocked the road from vehicular traffic and a normal takeoff was performed. The pilot flew directly to the Jerseyville airport, approximately 14 nautical miles from the point of takeoff. An additional 25 gallons of fuel were taken at the airport. The aircraft was flown the remaining 17 miles to the original destination without [further] incident. There were several factors involved with the causation of this incident: (1) winds aloft were significantly different than those forecast resulting in a lower than expected groundspeed, (2) a failure by the pilot to recognize the effect that the wind was having on fuel endurance and then to take positive action, (3) an overestimation of anticipated reserves, (4) trusting and/or improperly interpreting the fuel gauges. An incident of this type could be avoided by taking positive action when it is recognized that winds aloft and groundspeed are different than planned for and not relying on anticipated fuel reserves to complete a flight.

The pilot (who refers to himself throughout the report as "the pilot") wrote that he had calculated his true airspeed to be 118 knots over a trip of 280 nautical miles. He went on to say that this would give him, "4 hours and 4 minutes" of time en route. This calculation is incorrect. The numbers given would have yielded 2.4 hours. But the pilot makes some excellent recovery points at the end of his report. You should never completely trust the forecast winds. Doing so is one of the quickest ways to lose awareness of the actual situation.

The good news is that you do not have to trust the wind forecast. Once in flight you can calculate with great accuracy the actual wind's direction and velocity. If the actual wind is close to the forecast wind that you used when planning the flight, then stick to the plan. But if the winds are different you must be ready to change the plan in midflight. The pilot from the previous report failed to make changes in his plan. He became unaware of the actual situation and as a result placed himself in danger.

To determine the actual wind while in flight, the pilot must know five items. The pilot must know the true course, the true airspeed, the actual groundspeed, the wind correction error, and the true heading. All five of these items can be discovered in flight.

The true course and true airspeed should be known even before the flight during the preflight planning. The actual groundspeed is calculated in flight by taking the actual time between two checkpoints when the distance between the checkpoints is known. The last two items the actual wind correction angle (WCA) and the true heading (TH) are determined by actual trial.

To determine the actual wind correction angle, the pilot must be able to see a checkpoint up ahead. This is where it pays to select good checkpoints. The pilot looks ahead and while still about 6 to 8 miles away spots the next checkpoint. The best checkpoints for this are large smoke stacks, airports, lakes, or small towns. The pilot notices that the airplane's ground track is leading away from the checkpoint. In other words, if the present heading is maintained, the wind will blow the airplane to one side and the airplane will not cross the checkpoint. The pilot turns the airplane into the wind. This turn sets the wind correction angle. After a few minutes, the airplane is established in a "crab" angle that will cause the airplane to cross over the checkpoint. The true heading of the airplane is read from the compass or heading indicator. The wind correction angle is the difference between the true course and true heading. With the TAS, TC, GS, and WCA, the actual wind can be discovered using a standard flight computer. Follow these steps with the wind face side of a flight computer to determine the actual wind direction and velocity.

Step 1. Place the TC under the true index of the flight computer.

Step 2. Slide the movable card so that the GS is under the center hole.

Step 3. Mark a "wind dot" at the intersection of WCA and the TAS.

Step 4. Rotate the wheel so that the "wind dot" is under the true index.

Step 5. The actual wind direction is read under the true index and the actual wind velocity is read between the "wind dot" and the center hole.

Pilots should always be ready to replan their navigation numbers while in flight, because the atmosphere we fly in is ever changing, therefore the pilot must be ready to change with it.

ASRS NUMBER 441280 My weather briefing called for VFR conditions en route with marginal VFR over Athens, Georgia. The ceiling within 20 miles of Athens was 6000 to 7000 ft MSL. Athens itself had scattered clouds at 1300 feet broken and 4000 overcast. My altitude en route was 4500. Encountering the clouds at Athens I descended to 2500 feet MSL. After clearing the MVFR conditions around Athens, I decided to stay at 2500 feet. During the portion of my flight above 3000 feet I had leaned the mixture. I made a short stopover at Columbus, Georgia at which time I tried to contact FSS to extend my ETA at Pensacola. Attempts were to no avail. After departure out of

Columbus, my altitude was again 2500 feet. Upon reaching the outskirts of Pensacola approach radar, I started to contact Pensacola approach. I had the frequency dialed in when I picked up some engine roughness. Upon initial contact with approach, I notified the controller of my dilemma. He asked me if I would like to continue to my destination or land at a closer airport. I informed him that I would like to get the airplane on the ground as soon as possible. He told me to contact Eglin departure. After contacting Eglin, I lost my engine completely. I told Eglin that I was 20 miles out from Crestview. At 900 feet MSL I regained power and climbed to 2500 feet. During this time Eglin never reported radar contact. I had been issued a squawk code and tuned it in, but I was never picked up on radar. Even after squawking 7700 I was still not located. Eglin approach control asked me for a position update. I was on the 315 degree radial from Crestview VOR when I lost the engine for a second time. I located an airport and told Eglin that I had found an airfield and it looked as if I could make it to that runway. I told Eglin that there were T34s in the traffic pattern. It turned out to be Brewton, Alabama. I couldn't get in touch with the T34s and they were using the opposing runway 12, so I could not land on runway 30. I decided to set the airplane down in the grass infield instead of risking a head-on with a T34. I secured the aircraft, went through my emergency checklist, and did a soft field landing. No damage to the airplane, myself or my passenger. After landing I looked at the fuel gauge and it read $1/4$ tank. I got out and visually inspected the tanks for the third time that day. There was no fuel aboard except for the unusable fuel. I looked at the Hobbs meter and in 3.3 hours I had burned 40 gallons of usable fuel and 1.7 gallons of unusable fuel. I had visually checked the fuel before takeoff and they were topped. I visually checked the fuel at Columbus and the tanks appeared $1/2$ full after 2 hours of flight time. All Cessna 172s that I have flown all burned between 8 and 9 gallons per hour. This Cessna 172 had a 180 horsepower engine, so I used 10 gallons per hour burn rate for my flight planning calculations. This would have taken me to Pensacola with 30 minutes reserve for VFR daytime. The million dollar question: Why didn't I get more fuel at Columbus? After visually checking the tanks and "knowing" that I had plenty of fuel to get to my destination, I decided that I didn't need it! My million dollar question: Why wasn't I informed that this aircraft burned closed to 13 GPH? The operating handbook says 10. I believe that if the FBO at Columbus might not have been busy, I would have picked up some more fuel there, if for nothing else but to avoid the ramp fee. Like I have stated, it did not appear necessary. Lessons learned: (1) familiarize yourself with each individual aircraft when renting because the airplane may not perform as the POH indicates, (2) always get fuel when the wheels are on the ground regardless of the distance to the ultimate destination, (3) talk to mechanics who work on the airplane about pertinent information. The actions that I have taken since this incident to insure that this is not a repeated problem: (1) went over fuel calculations with an instructor, (2) received further instruction in cross-country planning, (3) I have given much more deliberation on my part about being more cautious and continue my flight training as an extremely safe conscious pilot.

Several references have been made to inaccurate fuel burn information in these past few reports. Very often pilots will find that the gallon-per-hour burn rate of a particular airplane will vary from what the performance charts report. The actual fuel burn rate does depend on several factors that can change from flight to flight. Air temperature, altitude, and leaning techniques will alter the GPH. The best solution is not to depend on the book at all. Calculate the GPH for the airplane yourself. If you go on a cross-country flight, leave with full tanks and refill the tanks after landing. Divide the gallons needed to return the tanks to full by the hours (and tenths of hours) of the flight. This will tell you the exact GPH of that airplane under those circumstances. Do not trust anything else.

The pilot in the last report blames a lack of knowledge on the *Pilot's Operating Handbook* and someone else "informing" him of the fuel burn. But he had a perfect opportunity to discover the actual GPH. He could have filled at Columbus and calculated the actual GPH. He then would have had a real number to base his fuel consumption calculations on for the remainder of the flight.

NTSB REPORT NUMBER ANC89FA119   The pilot reported that before reaching his destination, the engine lost power over a wooded and mountainous terrain. He turned the aircraft back to land downwind on a sandbar, but overshot and touched down in willows. He stated that earlier he had fully serviced the airplane and flown 1 hour. He calculated there were 4 hours of fuel remaining and the flight should have taken only 2 hours and 40 minutes. Draining of the fuel tanks after the accident revealed there was 5 gallons in the left wing tank but less than one pint in the right tank. According to the flight manual, this left only 1.5 gallons of usable fuel. The pilot reported there was light turbulence during the flight. Calculations revealed the aircraft needed an additional 5.4 gallons to complete the flight.

Probable Cause: Improper planning/decision making by the pilot, which resulted in fuel starvation. Factors related to the accident were: inaccurate fuel consumption calculations, inadequate fuel supply, turbulence, tailwind on landing, terrain conditions, and trees in the emergency landing area.

This was the pilot of a Cessna 207. He was a commercial pilot and a flight instructor. He was rated for both multiengine airplanes and helicopters. This means that a point of diversion can be encountered by experienced pilots as well as low-time pilots. This pilot could have and should have noticed what was going on, but he lost awareness of the fuel situation and ended up in the trees.

NTSB REPORT NUMBER BFO92LA136   The pilot was descending through 4000 feet en route to his destination which was 75 miles away when the left engine lost power. The pilot stated that he followed the emergency checklist and power was restored; however, shortly thereafter both engines lost power. The pilot radioed his destination approach control that he had a *sick passenger* on board, and asked for the closest airport. He was provided vectors to an airport 10 miles away, but decided to make a forced landing in a field as he stated he was low on fuel. According to the FAA, a visual inspection of both nacelle fuel tanks disclosed no evidence of fuel, and no fuel vapors were present. The examination also revealed that the tanks were not ruptured or damaged. The pilot reported that there was no mechanical malfunction, nor could the basis for the report of the sick passenger be established.

Probable Cause: The pilot's in-flight planning/decision making which resulted in fuel exhaustion.

This Piper Seminole pilot ran out of fuel because of improper planning but told ATC that he was going down to accommodate a "sick passenger." I guess that if I were sick in an airplane, I would rather be taken to an airport than to a farmer's field, but we say strange things when under stress. Fortunately, neither the pilot nor the sick passenger was hurt on landing.

NTSB REPORT NUMBER ANC98LA022 The pilot had departed from a remote off-airport area during a dark night. During the flight to the intended destination, a layer of clouds obscured the ground. The pilot continued in VFR on-top conditions and became lost. The pilot climbed to about 10,000 feet MSL to remain above the clouds. The airplane was equipped with a turn and bank indicator, but did not have any instrument panel lighting. The airplane's radio was not functioning effectively to establish contact with an FAA facility. The pilot was using a flashlight to illuminate the instrument panel, but the bulb burned out. The passenger in the airplane contacted an Air Route Traffic Control Center (ARTCC) via a cellular telephone and requested assistance in getting down through the clouds and locating an airport. The pilot, passenger, and FAA controllers worked together for about one hour in an attempt to assist the flight. About three hours after takeoff, the airplane's engine ran out of fuel, and the pilot began a descent through the clouds. The pilot reported that he attempted to maintain control of the airplane, but experienced two or three uncontrolled spins. As the airplane neared the ground, the passenger established visual contact with the ground and yelled to the pilot. The pilot then pulled back on the control stick, and the airplane struck the ground in an area of small hills. Rescue personnel located the airplane the following day. The pilot did not have an instrument rating. His pilot's certificate contained a limitation that prohibited night flying.

Probable Cause: The pilot's continued VFR over-the-top operation, becoming lost and disoriented, his inadequate in-flight planning/decision making, and the subsequent fuel exhaustion. Factors in the accident were the presence of a cloud layer below the pilot's selected altitude, the pilot's spatial disorientation, the pilot's lack of night flying experience, dark night conditions, and the lack of instrument panel illumination.

The pilot and passenger both escaped this flight with only minor injuries, but look back at all the places where the pilot lost awareness. It was almost as if the pilot was so removed from the flight that he was unaware of the consequences of his actions. The pilot took off at night even though he was not authorized to fly at night. The pilot flew at night even though the airplane had no instrument panel lights. He flew out over a cloud layer even though he was not instrument rated. He placed himself in a position where he would need assistance to get down, but the airplane's radios did not operate. The pilot flew around without resolving the problem until the fuel ran out. Thankfully (almost miraculously) the pilot lived through the ordeal that he put himself in. Let's hope he scared himself into better awareness.

NTSB REPORT NUMBER ATL89LA083 The pilot had made a flight from his home airport to Augusta, Georgia and upon returning, had to divert due to weather. He had departed Augusta about 2315, landed at Dekalb-Peachtree airport in Atlanta about 2400, and obtained a weather briefing from FSS about 0122. The briefer indicated that weather conditions were IFR and the pilot stated that he was "instrument" but the airplane had no attitude indicator. The pilot indicated that he departed Dekalb-Peachtree for McCollum about 0245. At 0306, Atlanta Center recorded a call from the pilot that he "got into some soup but was VFR." He indicated that he was not IFR qualified. After some vectors, the pilot indicated that he could see the ground. He was turned toward Dekalb-Peachtree at his request and then radioed that he was out of fuel. He had indicated sufficient fuel for 1.5 hours earlier in the sequence of events. Total time from initial radio contact to fuel exhaustion was about 32 minutes. During the forced landing the aircraft collided with a power line and trees.

Probable Cause: The pilot's decision to attempt VFR flight in adverse meteorological conditions at night with an inoperative attitude indicator and insufficient fuel reserve for the intended flight.

The private pilot of the Cessna 172 was the only occupant of the airplane and was seriously injured in the crash. This is a story that seems to repeat itself. A pilot, not instrument rated, flies into the clouds, then seeks help from ATC. Between cloud penetration and calling for help, there must have been some form of realization taking place—an awareness awakening. The pilot must not have thought flying in the clouds (this time at night) was too serious, but he quickly became aware that it was serious and called for assistance. Flying IFR is very challenging for IFR pilots, so it should follow that it could be deadly for VFR pilots. That is the reality. But this pilot's perception of reality was that flying IFR must not be that bad. This created a point of diversion during the flight that almost led to a tragedy.

Regaining awareness in time can prevent accidents. The following is the report of a flight crew who found themselves in a what would have been a critical fuel situation.

ASRS NUMBER 441301 We were under the control of Kansas City ATC and they gave us a rerouting for weather and en route spacing for Denver. The new route sent us way south of Alamosa and our Flight Management System together with our own calculations determined that with the new route we would not have enough fuel for Denver and Colorado Springs as an alternate. We asked for a better route but they told us that they were unable. ATC then asked if we were an emergency aircraft. We stated no. They stated that we could not proceed to Denver on a more direct route unless we declared a fuel emergency. The Captain stated that we were dispatched with enough fuel for a legal flight to Denver using Colorado Springs as the alternate. Then ATC gave us direct to Denver and declared us as an emergency aircraft. The rerouting would have brought us some 100 to 150 miles off our original course which most aircraft would not be able to comply with without stopping short of their original destination.

A point of diversion arrived during this flight (the route change), but the crew maintained their total awareness of the situation and refused the new route. Due to other traffic or ATC limitations, the only way that the crew could reject the new route was to have emergency authority to go straight to Denver. Then a tug of war followed over who would make an emergency fuel declaration. Ultimately the controller declared the airplane an emergency airplane so that he could approve the direct course. The crew would have been forced to land short of the destination otherwise. But the crew's awareness of the situation made it clear to them that they could not take the new route as is, and they immediately began working on a safe solution.

Weather Decisions

ASRS NUMBER 444119   I was flying home after an evening in beautiful weather. A severe storm came up and I tried to make it home. I got as far as a neighboring airport and their lights would not come on. It was becoming quite turbulent as the storm got closer. I had to turn and run. I was skirting Rochester Class C and trying to read my sectional chart for frequencies. The airplane is equipped with red lights only (it is a rental). It was impossible to read my sectional chart under the red light and in severe turbulence. The tower had been watching me and began signaling me with the light gun to come on in. I had been flying with full lights on, so I acknowledged his gun by blinking my landing lights and I came on in. I taxied to the FBO and called the tower. He told me I'd done great and everything was fine. I'm filing because the airplane owner advised it.

Something started taking place during this flight that the pilot was unaware of—the weather was changing rapidly. The pilot was out enjoying himself in "beautiful weather," but a point of diversion took place. The weather was becoming dangerous while the pilot thought it was beautiful. The weather can't be both beautiful and dangerous, so there was a reality gap. The pilot did not know what was really going on, so his response to the situation was delayed. The tower controller said that he had done great, but great to the controller means that you got down and no paperwork was necessary. In fact, the pilot did not do "great" because he did not maintain his situational awareness.

ASRS NUMBER 443404   I was southeast of Rhinelander, Wisconsin at 7500 feet VFR and was aware of a line of weather reported from Mason City, Iowa to Marquette, Michigan. Approximately 70 miles north of Green Bay I called FSS on 122.55 and requested a weather update. I was advised that VFR conditions should prevail on into Green Bay airport.

Approximately 60 miles north of Green Bay I encountered light rain and strong turbulence. I called FSS and reported disorientation and difficulty holding course and altitude. The FSS was extremely helpful and suggested that I squawk 7700. Then my radio went out and my VOR. This was caused by moisture coming in through the wing air vents. I had great difficulty reducing altitude due to rapidly rising air. Gradually I was able to get down to 5000 feet which placed me back under the weather, into VFR conditions, and out of the developing thunderstorm. I had a portable GPS and indicated that the nearest airport was at Marinette/Menominee, Michigan which was approximately 15 miles away. I landed without radio assistance. I phoned Green Bay FSS at once and reported the conclusion of the emergency flight. The radios have since been repaired and I have enrolled in a program for additional instrument training.

The pilot approached a developing storm but believed that "VFR conditions should prevail." What was real and what he thought was real were different. The pilot did take action once reality did become apparent by descending and diverting.

NTSB REPORT NUMBER ANC95LA020   The air taxi pilot reported he departed St. Mary's Airport with one passenger aboard. He said he was flying about 600 feet above the ground along the south bank of the Yukon River, when opposite direction traffic reported inbound to St. Mary's, also flying along the south bank. The pilot said that he elected to transition to the north bank to avoid the traffic, but soon encountered deteriorating weather conditions. While trying to keep visual contact with the shoreline and trees, he said he inadvertently allowed the airplane to descend until it struck snow covered terrain.

Probable Cause: The pilot's decision to continue visual flight into instrument meteorological conditions. Factors associated with the accident are the pilot's failure to maintain adequate altitude/clearance from terrain, and the whiteout weather conditions.

The poor visibility, the incoming traffic, the snow, and white-out conditions were all factors, but where an awareness of safety first departed this pilot was when he elected to fly VFR with a low overcast, forcing a cruise altitude of only 600 feet.

NTSB REPORT NUMBER ANC92FA116 The destination camp is located about 50 miles southwest of the departure lodge, and is separated by mountains with some peaks over 4000 feet MSL. The general area weather included low ceilings. The operator stated that shortly before the accident he observed the Youth Creek weather "better than 400 feet...", and told the pilot that it appeared good enough to make the flight. The pilot stated that as he flew up the creek he could see that the pass was closed by low clouds, and attempted to reverse direction. The pilot stated that he "made a hard, steep turn, and the airplane stalled...." The pilot said the weather was about 800 foot ceiling and visibility of one mile. When the pilot was admitted into the hospital immediately following the accident, he was diagnosed as having an insulin dependent diabetic condition. His medical records contained no evidence of, and the pilot denied any knowledge of, any pre-accident diabetic condition.

Probable Cause: The pilot delayed decision in reversing course and his failure to maintain airspeed during the maneuver. Factors related to the accident were: mountainous terrain and low ceiling.

Had the pilot been hiding a medical condition? Maybe, but what led to the accident was the pilot's decision about the low clouds. He heard that the ceiling was "better than 400 feet," so he concluded it was good enough to go. In reality it is never "good to go" VFR when the ceiling is only 400 feet.

NTSB REPORT NUMBER BFO95FA020 The noninstrument rated private pilot was on a cross-country flight at night over a mountainous area. During the flight, he contacted ATC and requested flight following services. ATC advised him to maintain

5500 feet MSL, and he climbed to that altitude. During a frequency change (from Wilkes Barre approach control to Allentown approach control) radio contact with the airplane was lost. The controller tried several times to reestablish radio contact, but without success. The airplane subsequently collided with trees on top of Broad Mountain and was destroyed on impact and fire. Wreckage was scattered over a 600 foot area. The pilot had received a weather briefing that indicated marginal VFR weather and flight precautions for mountain obscurations along his route of flight. About 25 miles south at Allentown, a special weather observation at 1822 EST reported snow showers with $\frac{1}{2}$ mile visibility in fog. Examination of the wreckage did not disclose evidence of mechanical malfunction.

Probable Cause: The noninstrument rated pilot's poor in-flight decision to continue VFR flight into instrument meteorological conditions, at night, and his failure to maintain sufficient altitude to ensure clearance from obstructions/terrain. Related factors were the instrument meteorological flight conditions (low ceiling and fog), the dark night, and the pilot's lack of total experience.

A fatal point of diversion took place when the hand-off between controllers was not accomplished. The pilot apparently did not realize that he was approaching higher terrain. If the handoff had been made and the pilot had been talking to the next controller, that controller would have asked him to climb. But the pilot remained unaware of the problem until it was too late. The pilot and one passenger were killed in the accident.

NTSB REPORT NUMBER ATL98FA023 Following a missed approach at the destination, the pilot requested weather information for two nearby airports. One airport was 53 miles northeast with a cloud ceiling of 900 feet, and a visibility of 6 miles. The pilot opted for the accident airport, 21 miles southwest, with an indefinite ceiling of zero, and visibility of $\frac{1}{4}$ mile. After completing the second missed approach, the flight proceeded to the accident airport. Radar vectors were provided to the ILS runway 36L (Charlotte Douglas International Airport). On the final approach, the flight veered to the right of the localizer and descended abruptly. The last recorded altitude for the flight was below the Decision Height. Investigation revealed no anomalies with the airport navigational aids for the approach, and the airplane's navigation receivers were found to be operational. Postmortem examination of the pilot did not reveal any preexisting diseases, and toxicological examinations were negative for alcohol or other drugs.

Probable Cause: The pilot's continued approach below Decision Height without reference to the runway environment, and his failure to execute a missed approach.

The decision to divert to an alternate airport, alternate altitude, or alternate route is always a test of awareness. This pilot made a poor diversion decision because he passed up an airport with a 900-foot ceiling for an airport with a zero foot, indefinite ceiling. The next report is from a pilot who made a better diversion decision— but it was a close call.

ASRS NUMBER 441890   We declared "emergency fuel" during an unplanned diversion to Las Vegas, Nevada (LAS). The flight was required to divert to LAS due to unforecast dust storm rapidly and dramatically reducing the visibility at Phoenix (PHX). The flight released with scheduled fuel load of 11,000 pounds, which included the standard reserve of 4000 pounds plus 1200 pounds of "extra fuel." Current and forecast weather for Phoenix was VFR conditions with no hazards noted. We obtained ATIS just as we were to begin our descent. The weather reported calm winds, 10 miles visibility. As the descent continued the approach controller began issuing airspeed adjustments and delay vectors due for a dust storm north of the airport. We discussed the situation with our dispatcher on our company radio frequency. Approximately 5 minutes later the situation had further deteriorated. We were advised that we were number 12 in the approach sequence and that Phoenix had become IFR from blowing dust. As a new Captain on the

A320 my personal minimums were 1 mile or 4500 RVR. Visibility was dropping rapidly. The dispatcher and I agreed a diversion back to LAS was in order. Upon reaching our return cruise altitude, I first declared "minimum fuel" and then to ensure priority handling I declared "emergency fuel." Our onboard computer computed a landing fuel of 1200 pounds at LAS. An uneventful approach and landing was made at LAS. We shut down at the gate with 1060 pounds of fuel.

The pilot and crew of this flight did a good job of recognizing the problems at hand and maintaining awareness. They processed information well. They used information from ATIS, the controllers, and their own dispatch. When they were told they were number 12 in line, they made the command decision to divert. The captain made all the right moves—first declaring minimum fuel then emergency fuel. Their awareness fostered good decisions and they made it—but with only a few minutes to spare. Any lack of awareness or any indecision would have led to an accident.

Aircraft Operations

The last set of reports comes from the common operations of aircraft. They involve takeoff, en route, approach, and landing situations. The points of diversion are present in every case.

Takeoff decisions

NTSB REPORT NUMBER ATL91LA156   The student pilot was performing the takeoff of the return leg on a VFR cross-country flight. At about 55 to 60 knots, he began to rotate, and the stall warning horn activated. The airplane was not reaching flying airspeed, so he reduced the throttle and applied brakes. He was unable to bring the airplane to a stop in the remaining runway surface, and the airplane collided with a tree and a creek bed. The pilot reported no mechanical problems with the airplane, and none were found during the postaccident inspection. The pilot, whose total flying time was 21 hours, reported the temperature at the takeoff to be about 100 ° F.

Probable Cause: The student pilot's delayed decision to abort the takeoff, resulting in his inability to stop the airplane in the remaining runway. Factors were the pilot's lack of total flying experience, and the high outside air temperature at the time of the accident.

The student pilot, flying this Grumman AA-5B, was unhurt after hitting the trees. The student must not have considered the 100-degree temperature and the density altitude effects. The student was not aware that a longer takeoff roll and a faster liftoff speed would be required under these conditions.

NTSB REPORT NUMBER ATL91LA053   The pilot stated that during takeoff, the airplane became airborne, but would not climb or gain airspeed. Subsequently, it was damaged during a forced landing. No preaccident mechanical problems were reported. An investigation revealed that the aircraft was approximately 175 pounds over its maximum allowable gross weight. The fuel on board the aircraft was gold in color and smelled like auto fuel.

Probable Cause: The pilot's improper planning/decision, and his attempt to fly the aircraft while it was loaded over its maximum allowable gross weight.

This pilot thought he could take off normally even though the airplane was significantly overweight. What the airplane could do and what the pilot thought the airplane could do were different. The pilot missed a point of diversion and an accident resulted.

NTSB REPORT NUMBER ANC92LA140   The pilot stated that after getting airborne the airplane would not climb any higher than about 2 feet above the ground without the stall warning horn sounding. About 20 feet past the end of the runway the airplane's right main landing gear was hit and was sheared off by a 3 foot high berm. The pilot then reduced power and landed on a paralleling beach, during which the nose landing gear collapsed. The pilot also said that the airplane was delayed in getting airborne by the standing water and mud puddles on the runway.

Probable Cause: The pilot's improper preflight planning/decision, and his delayed aborting of the takeoff. Contributing factors were the berm and the standing water on the runway.

Five people were aboard this Piper Navajo, but no one was hurt in the accident. Mud and standing water slowed the airplane's acceleration to a safe takeoff speed.

NTSB REPORT NUMBER ANC96LA024   The pilot was attempting to take off and return to Fairbanks, Alaska. He had to taxi uphill for a distance of 1500 feet. He used carburetor heat during his taxi. When he turned the airplane around to line up for takeoff, he noticed fine snow and ice crystals suspended in the air over his intended takeoff area. The pilot turned the carburetor heat off, applied full power, and made the takeoff. When the airplane reached approximately 15 feet above the ground, the engine lost power. The pilot applied carburetor heat and some power was restored. The pilot stated he had to lower the nose to maintain airspeed and the airplane crashed into the trees. The pilot stated he thought the intake screen had become iced during the taxi and takeoff.

Probable Cause: The icing of the induction/intake screen and the pilot's improper planning/decision by electing to take off into suspended snow and ice crystals.

The pilot was unaware that the air intake screen could become completely closed by the ice. He saw the "fine snow and ice crystals" in the air but didn't think it was a problem. The pilot was unhurt, but his Cessna 180 was damaged.

NTSB REPORT NUMBER ANC89LA172   The pilot was on a flight to airdrop survival equipment to survivors of a boating accident, of which he had witnessed the drowning of a close friend. He departed in gusty wind conditions. The rescue area was located on the lee side (downwind) of mountains that rose to about 2000 feet above the lake. According to witnesses, the aircraft was maneuvering at about 150 feet AGL, before the accident occurred. The pilot reported that the aircraft encountered severe turbulence and a down-draft from which he was unable to recover. Witnesses estimated the wind was from 120 degrees at 35 gusting to 60 knots. The pilot estimated the wind was gusting to 80 knots.

Probable Cause: Improper planning/decision by the pilot. Factors related to the accident were: the pilot's self-induced pressure and the adverse weather conditions.

Rescue work brings in a whole new element and a new challenge for maintaining awareness. A pilot "on a mission" will sometimes turn a blind eye to an obvious safety problem. Pilots might fly in conditions that otherwise they would not, but they have emotions running high. The pilot in this accident was obviously emotional—having had a close friend killed in another accident. He probably would not have flown in wind gusting to 80 knots unless he felt he had a strong overriding reason. But even in these situations, pilots must remain objective and try to maintain awareness without being influenced by outside pressures.

I had a former student who was a medical emergency rescue helicopter pilot, and he asked me to help him design a proposal to help rescue pilots maintain a clear awareness. The problem was that most rescue pilots are so mission-focused that they will take chances to complete the rescue. These pilots have a real "comrade down" attitude. If they know that someone has been in a car accident and will die if they don't get their helicopter in and out, they will ignore obvious safety precautions and accept big risks. In the 1980s there was a rise in rescue accidents that were attributed to this attitude. To help these pilots focus more on safety and less on mission, our proposal included an after-takeoff notification. The pilot would have to take off and then when airborne they would be told if the flight was for an actual rescue or for a drill. When the pilot knew it was a real rescue before takeoff, he would look at bad weather and sometimes say—let's go anyway. But under the proposal, when a call came in, the pilot knew that it might be a drill and he would not want to risk flight in bad weather just for a drill. This forced the pilot to make go/no-go decisions based on safety considerations alone.

Then my former student was himself killed in a rescue accident, although it did not involve poor weather decisions. The next report is of the East Care accident in 1987. One of the medical oxygen canisters carried on the helicopter for patients to breath en route to the hospital exploded for still unknown reasons.

NTSB REPORT NUMBER ATL87MA057 The emergency medical evacuation helicopter was on a night flight with a 3-month old patient aboard. After reporting level at 3000 feet, the pilot transmitted that he was going to make an emergency landing. This was followed by several unreadable transmissions that approach control personnel thought were in reference to an onboard fire. At about the same time, a flight nurse on the helicopter transmitted on the hospital frequency that they were on fire and were going down. Soon thereafter, radar and radio contact were lost. Subsequently, the helicopter crashed in a slight nose down, right bank attitude and burned. It was extensively damaged by impact and fire. Crash damage revealed the helicopter had impacted at a high vertical velocity and with little or no forward movement. No preimpact part failure or malfunction of the engine or airframe was found. The helicopter had been modified for emergency medical service operations in accordance with 4 supplemental type certificates (STCs). The modifications included installation of a high pressure oxygen (O_2) system. During the accident, several oxygen lines failed from impact or fire, which created a blowtorch effect. There was no provision for emergency shut-off of the oxygen at or near the O_2 cylinders. The original source of fire was not found.

Probable Cause: Reason for occurrence undetermined.

In-flight decisions

NTSB REPORT NUMBER ANC89FA090   The pilot received a weather briefing earlier in the day. During departure the pilot requested Stage II radar services to the south. The aircraft never arrived at its destination. Five weeks later the aircraft was located where it had crashed at an elevation of about 7000 feet, near the top of a mountain ridge. Impact occurred inside a box canyon on steep, snow-covered terrain. An exam of the wreckage indicated the landing gear was retracted during impact. The propeller was not found, but 4 propeller attach bolts were found that had been sheared. Approximately 37 miles away and at an elevation of 1294 feet an airport reported weather as 5000 scattered, 12,000 scattered, 18,000 thin broken, visibility 35 miles at the time of the accident.

Probable Cause: Improper in-flight planning/decision by the pilot, and his failure to maintain sufficient altitude while entering a box canyon and or crossing mountainous terrain in restricted weather conditions. The terrain and weather were related factors.

This happened during daylight hours. It is possible that the flight entered the canyon in VFR conditions, but the pilot was unaware of the space needed to escape the canyon.

NTSB REPORT NUMBER ATL92LA156   The pilot reported that immediately after takeoff, he had a collision with a flight of birds. He stated that the right engine immediately had a partial loss of power. He stated that he did not attempt to raise the landing gear nor the flaps following the loss of engine power, and the aircraft would not maintain altitude. Examination of the aircraft engines revealed that there was rotational scratching of the turbine housing on the right engine, and no rotational scratching on the turbine housing of the left engine.

Probable Cause: The poor in-flight decision by the pilot in command in that after a partial loss of power, he shut down the wrong engine, did not follow the emergency checklist, did not raise the flaps, and did not retract the landing gear. A factor in the accident was the in-flight collision with birds during the takeoff climb.

One of the keys to maintaining situational awareness is to have a high degree of technical proficiency. This pilot did not perform the basic engine-out emergency procedures and therefore turned an instant of excitement (when the birds hit) into an accident. The pilot was seriously injured and the airplane was substantially damaged after the pilot shut down the wrong engine.

ASRS NUMBER 441739 While I was on final approach, the fuel starved from the left tank. I was unable to switch to the right tank soon enough for a restart. I made a precautionary landing on highway 93, which is 1 mile south of Jackpot, Nevada. There was no damage to the aircraft or to autos or to other property.

This was the pilot of a home-built Long Ez. The fuel was running out of one tank and the pilot was not aware of it. The pilot only regained awareness of a fuel problem when the engine stopped.

NTSB REPORT NUMBER ANC93LA011 A student pilot under dual instruction and instructor on a cross-country flight completed a cruise descent with low engine power and developed a rough running engine. The instructor took control and efforts to regain power with carburetor heat and full throttle were ineffective. Attempts to position for an emergency landing at strip were not successful due to high winds. An alternate landing area was selected on a dry river bed, however, the decision to turn into the wind was not made until approximately 200 foot AGL, and too late to avoid a tailwind landing. During the tailwind landing the airplane's nose wheel was destroyed on initial impact and the airplane nosed over.

Probable Cause: The pilot in command's (flight instructor) in flight planning/decision to attempt a downwind forced landing in high surface wind conditions and the pilot's improper use of carburetor heat.

Ice started to form in the carburetor, but the student and flight instructor did not realize it. The conditions were favorable for carburetor ice so they could have been aware of the possibility. Neither the student nor the instructor were hurt after they made an unwise tailwind landing in this Cessna 150.

Approach to land

ASRS NUMBER 449839   The incident occurred while we were holding short of runway 35 at Valdosta. We were a scheduled, part 121 flight to Atlanta. We taxied out for departure and were advised by the tower that they were coordinating with Jacksonville Center for our clearance. This was a Sunday and as a result Valdosta Approach Control was closed so Jacksonville Center was issuing all arriving aircraft the full ILS runway 35 approach into Valdosta. This approach requires the pilot to fly direct to the VOR that is located on the airport, then fly outbound on the 212 degree radial, then join a 12 DME arc to the localizer. Valdosta tower advised us of an inbound Bonanza on this approach and that we would be released after he landed. The Bonanza proceeded to the VOR as published and began to track outbound on the 212 degree radial. The aircraft we were in had a "look up" feature on the TCASII that allowed us to "follow" the Bonanza as he began the approach. The weather was 1300 feet overcast, with 5 to 7 miles visibility. As the Bonanza joined the 12 DME arc we were no longer able to watch the aircraft on the TCAS as the "look up" feature only extends out 10 miles. Ten minutes passed and the Bonanza was not seen. About this time the Bonanza reported to the tower that he had all flags on both his glide slope and localizer. The tower advised that everything was operational and to check his radios. The Bonanza did that and then advised the tower that he was just going to track inbound back to the VOR. I would have expected the "execute missed approach" order at the same time, but instead he was told, "Roger, report the field in sight." The weather at the time was below any Minimum Safe Altitude (MSA) for vectoring and below any MDA. The Bonanza reappeared on our TCAS approximately 10 minutes later, 10 miles west of the field. The proper approach course would have put him 10 miles south. I questioned the tower controller about why the Bonanza was west and not south. With this the tower controller seemed confused and queried the Bonanza pilot. The Bonanza had now dropped below the cloud bases, reported the field in sight, and landed. Problems: (1) Because the approach control was closed on Sunday there was no radar service provided. Valdosta just happens to be at an elevation of 200 feet and the land is flat. If any other terrain existed, this Bonanza pilot and any passengers would have been killed. The poor weather conditions combined with a complicated approach, a "weekend" pilot, and an inoperative (closed) approach control facility easily combined to create a dangerous situation. It was the perfect set-up for a CFIT (controlled flight into the terrain) incident. (2) The controllers should have told the pilot to execute an immediate missed approach, but instead tried to "help" get this guy on the ground so that we could depart. The tower controller completely dropped the ball. (3) The Bonanza pilot should have recognized and admitted to the tower that he was lost and disoriented. As I see it this was no small FAR violation. It was many that could have quickly combined to be fatal.

The Collision Avoidance System (TCAS) that was on the commuter aircraft had a view of the Bonanza's approach that the controllers did not have. The Bonanza pilot was extremely fortunate not to have been in an accident after becoming disoriented. It seems that from the description given, the Bonanza pilot must have turned the wrong way on the DME arc. Instrument flight requires a very high degree of awareness. Part of maintaining that awareness is the ability to "see" yourself throughout an instrument approach. This Bonanza pilot did not do that and got lost in the procedure.

ASRS NUMBER 446714   I made a missed approach at my destination airport. I was then vectored to a second straight in NDB approach. I crossed the NDB at 1500 feet, and then dropped to 1300 feet. I then mistook a road for the runway and proceeded lower. I realized my mistake, but I became confused, circled under the ceiling and below the MDA. I found a different airport (Essex County) and got approval from New York Approach to land there. I contacted the Essex Tower and was cleared to land on runway 4.

The point of reality diversion came when the pilot saw what he thought was the runway and ended up following a road instead. The road lead him to another airport! There was no terrain or obstruction protection for this pilot when he flew under the clouds below the MDA to the other airport, so he was very lucky this was not an accident.

NTSB REPORT NUMBER BFO92FA031   The pilots were on the first leg of a personal flight. The pilot in command had obtained a weather briefing on the previous night, but had not updated it on the day of the flight. The weather at the destination airport was in part: 100 feet overcast, reduced visibility with fog, wind from 020 degrees at 17 knots. On the first approach, the pilot executed a missed approach, but did not follow the missed approach procedure. The missed approach procedure was to climb to 1000 feet MSL on the runway heading, then turn left while continuing to climb to 2800 feet via the localizer course to the outer marker. Instead, he circled to the right over the city and requested another approach. On the second approach, he flew to the left of the localizer and descended below the glide slope and decision height. The aircraft impacted trees on a hill about 200 feet below the ILS decision height and 1.85 miles from the approach end of runway 23 (Frederick, Maryland), A post accident investigation revealed no evidence of a preexisting mechanical failure.

Probable Cause: The pilot's improper IFR operation and descent below the ILS glide slope and decision height, which resulted in a collision with trees. The weather conditions and lack of a recent weather briefing were related factors.

The first point of diversion that started this fatal chain took place the night before as the pilot slept. He got a weather briefing the night before the accident, but never again made himself aware of what was taking place at the time of the flight. This meant that he took off in a state of being unaware and this, together with extremely low ceilings, sealed the fate of the flight. Both the pilot and the one passenger were killed.

Landing

NTSB REPORT NUMBER ATL98LA110 According to the pilot, he was 2.5 hours into a 4-hour flight when he felt an urgent need to relieve himself. The pilot stated he decided to land on a road in a cultivated field. The pilot stated that after landing, he noticed a post on the left side of the road and maneuvered to miss the post. During this maneuver, the airplane became airborne, and when it touched down a second time, the landing gear collapsed. According to the FAA inspector, the pilot stated he had to relieve himself, so he decided to land on an access road because he didn't see an airport. The pilot then stated that he didn't have enough clearance on the road, and decided to land in the field. The FAA inspector also stated the field was approximately 1 to 2 miles south of the Thomasville Airport in Thomasville, Georgia. According to the FAA. the airplane touched down on the edge of the field, crossed the access road, and came to rest in another field. When crossing the road, the right main landing gear was sheared off. As the airplane continued to roll, the lower third of the rudder and the fuselage were bent.

Probable Cause: The pilot's poor in-flight decision to attempt a forced landing in a field 2 miles from an airport, leading to an on-ground collision with rough terrain. A factor was the pilot's physiological need.

The pilot must not have been aware that an airport was only 2 miles away. He thought an off-airport landing was his best option, but he would not have thought that if he had known about the airport being so close. The pilot therefore must not have been aware of his position before he decided on the landing. Better awareness leading up to this event would have meant a runway and a restroom instead of an accident and a field. The pilot was otherwise unhurt.

NTSB REPORT NUMBER BFO93LA028   The pilot tried to activate the radio-controlled runway boundary lights about 10 miles away, and also while in the traffic pattern, but he was unsuccessful. He continued his descent to approximately 500 feet above the ground using the VASI (Visual Approach Slope Indicator) lights. The pilot stated that he had the airport in sight and "felt well enough in sight to complete landing." The airplane touched down in approximately 18 inches of snow 60 feet off the right side of the runway. The pilot reported that there was no mechanical malfunction. He said that as he got closer to the ground he realized it was snowmobile tracks and not the runway that he had seen. He tried to go around but the airplane impacted the ground collapsing the nose gear.

Probable Cause: The pilot's inadequate in-flight decision to continue a landing without runway lights, and his delay in initiating a go-around. A related factor was the pilot's over-confidence in his ability.

Landing at night without runway lights is dangerous, but the pilot did not think so. He lined up to the side of the VASI lights, but the runway was on the other side. Here we can see a clear gap between what was real and what the pilot thought was real.

NTSB REPORT NUMBER ATL96LA050   The pilot reported that about 40 miles from his destination, the airplane's lights blinked on and off due to a short in the system. He elected to land at Goldsboro, North Carolina, already aware of the fact that the runway lights at Goldsboro were inoperative. The airplane landed about 20 feet left of the runway, in grass, and collided with a piece of cultivating equipment. During examination of the wreckage, the landing light was operated several times and it functioned normally.

Probable Cause: The pilot's improper decision in attempting a night landing at an airport without operating runway lights that resulted in a touchdown off the runway. Factors that contributed to the accident were: An airplane lighting malfunction for an undetermined reason(s) and the dark night light conditions.

The first point of diversion took place when the pilot overreacted to his in-flight electrical problem. It was dark, but the weather was good VFR. The pilot could have chosen to fly to another airport where the runway lights were on. The pilot's original destination was only another 20 miles away and had 7500 feet of lighted runway.

Of course, reading an accident report and identifying what a pilot should have done is easier than identifying diversions in your own flight. The next chapter will help pilots take the last step. We must be our own best critics. We know that a lack of awareness could slip into our next flight and place us in jeopardy. Therefore we must prepare to overcome the problem. We must prepare to be aware.

6

The Penalty

Unfortunately, the penalty for the lack of situational awareness is accidents. The loss of awareness sets in motion a potentially fatal chain of events. First, factors develop that the pilot is not aware of. This leads to flawed decision making. Poor decisions produce poor choices, and poor choices create accidents—sometimes deadly accidents. The chain, then, is:

1. The loss of awareness.
2. Flawed decisions.
3. Poor choices.
4. Accident.

The link between awareness and the end result, accidents, is not always clear in accident reports. The accident investigator must recreate the sequence of events from little or no remaining evidence and then work back up the chain from accident to initial loss of awareness. Sometimes there simply is not enough information to make this link. For this reason, accidents that take place at the end of a chain that began with the loss of awareness are underreported.

When an accident investigation is concluded, a National Transportation Safety Board (NTSB) Factual Report (form 6120.4) is filed. The report includes as much information about the accident as possible. This includes the number of any people hurt or killed, the aircraft type, the location of the accident, and weather conditions. The report contains a narrative so that the investigator can explain what the search has discovered about the accident, and then finally the investigator makes a statement of "probable cause." Investigators never give an "absolute cause" statement because they allow for the fact that there may be factors not in evidence that could change their opinion. But based on what they do know, they make a statement of what they think "probably" happened.

I searched the NTSB forms from 1983 to 2000, over 40,000 reports, and only in 31 cases was the term "situational awareness" or "awareness" used as part of the probable cause statement. Only 31 references out of over 40,000 reports could make us believe that this was a very small problem. But in this case the lack of references does not tell the whole story. The term "situational awareness" is a phrase that has only been applied to aviation recently. Only NTSB reports from the mid- to late 1990s used the term. There have been many accidents that were clearly the result of a pilot's lack of awareness, but it was called something else prior to the term's general use in aviation. Second, it is sometimes difficult for investigators to piece together the entire puzzle after the fact sufficiently enough to call it a lack of awareness. The ability to move backwards in time and backwards up the chain from accident to the first occurrence of the loss of awareness is not easy. There are times when investigators cannot make the link and stop short by calling the accident the result of "poor planning" or an "improper decision."

The probable cause statements that do include a reference to awareness are similar, and each provides a picture of a situation that quickly got out of control. Here are some excerpts for probable cause statements taken from the NTSB reports:

> "The pilot's failure to remain aware of the local wind conditions, executing an approach with excessive speed, combined with a tailwind, which resulted in the runway overrun."

> "One factor leading to the midair collision was the failure of the pilot of the Cessna 172 to become aware of and use the listed frequency for interaircraft communications."

> "Before takeoff the pilot was aware of his fuel endurance, [but the accident occurred] due to the pilot's poor in-flight planning, in that he diverted his attention and continued the flight beyond the original period planned."

> "Fuel exhaustion due to the student pilot's lack of fuel awareness."

> "The Captain's decision to initiate visual flight into an area of known mountainous terrain and low ceilings and the failure of the flightcrew to maintain awareness of their proximity to the terrain."

> "The pilot's failure to maintain a proper glide path while on short final approach. The dark night conditions may have affected the pilot's awareness of the airport environment."

> "The Captain's actions that led to a breakdown in crew coordination and the loss of altitude awareness."

"The pilot's lack of situational awareness (becoming lost or disoriented during the instrument approach), his failure to fly the approach as charted, and his failure to maintain aircraft control."

"Improper use of carburetor heat which resulted in a power loss due to carburetor ice over unsuitable terrain. Factors related to the accident was the failure of the pilot to be aware of the temperature/dew point spread."

"The instructor pilot's loss of altitude awareness and possible spatial disorientation, which resulted in the loss of airplane control at an altitude too low for recovery."

All of these probable cause statements used the term awareness or situational awareness in them, and it was clear in each case that as soon as the loss of awareness took place, bad things were going to happen.

Because of the underreporting, the number of accidents that were specifically blamed on the lack of awareness as the primary cause was very low, so I looked to the next step in the chain—poor decision making. If the assumption is made that a loss of awareness leads to poor decisions, then a listing of accidents that were attributed to poor pilot decisions might indirectly yield information on awareness as well.

The NTSB files include accident reports from 1983 to the present. In addition there are two other accidents included in the files that took place prior to 1983 but were only discovered after 1983. One of these was an accident that took place on October 24, 1948, but the whereabouts of the wreckage remained a mystery for almost 40 years—finally being discovered in 1987. The other was an accident on July 19, 1962 that went unsolved until August 8, 1994. These two combined with over 40,000 others were used to seek out "flawed decision" accidents. The search revealed 1607 accidents that specifically attributed the probable cause of the accident to a pilot's poor decision making. Of these, 477 were fatal accidents. There are no specific guidelines for the language used in the NTSB reports. These 1607 accidents had a specific reference to faulty pilot decision making. It should be assumed that other accidents involving some form of pilot decision error exist, but do not specifically mention the error in the probable cause statement. These accidents would not be among the 1607. For this reason, I believe that the 1607 is also an underreport and that these numbers represent just a portion of the larger problem. Nevertheless, the numbers that are available are helpful in identifying trends.

A closer look at these probable cause statements gives us an idea of the pilot's state of awareness. These are excerpts from NTSB reports citing poor decision making as the probable cause:

> "The pilot's decision to commit himself to a landing before completing an evaluation of the weather in the vicinity of the landing site."
>
> "The pilot's improper planning/decision concerning fuel management..."
>
> "Improper in-flight planning/decision by the pilot and his failure to maintain adequate airspeed."
>
> "The pilot's overconfidence in himself and the aircraft..."
>
> "Unfavorable wind conditions and the pilot's self-induced pressure."
>
> "Contributing factors of the accidents were the student's overconfidence in his personal

abilities and his selection of the wrong runway, and resulting tailwind conditions."

"The pilot's continued flight into known adverse weather and his delay in reversing course..."

"The pilot's decision to take off downwind..."

"The pilot's failure in not identifying weather conditions along the route..."

"The pilot's failure to attain a proper climb. Factors in the accident were: the pilot's lack of familiarity with the airplane, his inadequate training in that model airplane, his lack of experience, his in-flight decision making, and his overconfidence in his personal ability."

"The pilot's delayed decision to abort the takeoff..."

"The pilot's improper in-flight decision and his failure to perform a go-around..."

"The pilot's improper in-flight planning/ decision, and his failure to maintain a proper altitude while operating in the mountains."

In each of these cases, if the pilot had been more aware of the wind, or the remaining fuel, or the length of runway, or the weather ahead, they might not have made such a poor decision. They could have avoided the accident with better awareness. I believe that poor-decision accidents are the same as lack of awareness accidents. If my belief is true, then we can analyze the 1607 accidents and learn where the greatest dangers from the lack of awareness exist.

Of these 1607 "poor-decision" accidents, 76 involved student pilots, 740 were private pilots, 655 were commercial pilots, 318 were single-engine flight instructors, and 166 were multiengine flight instructors. Note: The totals do not always add up to 1607 because there is some duplication among pilots. Although private pilots were involved in the most accidents caused by poor decisions (46%), commercial pilots also had a high number of flawed-decision accidents (41%). Since commercial pilots must have greater experience than private pilots, it would seem logical that commercial pilots would have far fewer decision-related accidents. What impact does experience in the form of flight time have on decision accidents?

The table below tells us the total flight time the pilot had when he or she became involved in an accident that had poor decision making listed as the probable cause:

0–100	131
101–200	141
201–300	105
301–400	104
401–500	76
501–1000	230
1001+	820

There is a slight reduction as pilots acquire from between 100 and 500 flight hours. Inexperience seems to be a factor in poor decision making, but over half of the accidents involved pilots with greater than 1000 flight hours. Maybe the factor that has a bigger impact on safety is not total time but how recent the time is. Sixty-nine percent of the accidents involved a pilot who had from 0 to 25 hours within the 90 days before the accident. Forty percent of the accidents involved a pilot who had from 26 to 50 hours within the 90 days before the accident. Only 1 percent of the accidents involved pilots who had flown between 51 and 75 hours within 90 days of the accident. The numbers can be further broken down:

Flight hours within the last 30 days before the accident.

0–10	38%
11–20	21%

21–30	14%
31–40	8%
41–50	5%
51–60	4%
61–70	3%
71–80	3%
81–90	2%
91–100	2%

Flight hours within the last 24 hours before the accident.

0–5	83%
6–10	16%
11–15	0.5%
16–20	0.5%
21–24	0%

Number of months prior to the accident that the pilot completed a flight review.

0–6	58%
7–12	21%
13–18	13%
19–24	8%

These numbers do reveal trends. Pilots have more poor decision-related accidents when they have not practiced often. It is true: Flying is not like riding a bike. You cannot just pick it up after a period of inactivity and begin again where you left off. Piloting skills require constant practice. In 198 (roughly 12%) of these accidents, the pilot was not even legal to be the pilot in command. Eighty-four pilots were flying without a current biannual flight review, and another 114 without a current medical certificate. These pilots' first flawed decision was to pilot the airplane in the first place—is it any wonder they were involved in an accident caused by poor decisions?

Of the poor-decision accidents, 296 took place inside instrument weather conditions. Of these 296 accidents, 103 of the pilots involved were not instrument rated. These VFR pilots decided to press on into deteriorating weather and found themselves in IFR weather—most all of these were fatal.

Not surprisingly, of the total decision-related accidents, 993 of them took place on what was determined by the NTSB to be "personal" flight. Another 160 decision accidents took place on "instructional" flights.

Is there a point in a flight that requires the greatest awareness and the best sequence of decisions? From the following table we can see that accidents are spread out across the entire flight, but descent, approach, and landing made up 63% of the accidents that reported a particular phase of flight.

Phase of Flight

Descent	9%
Approach	12%
Landing	42%
Takeoff	19%
Climb	4%
Cruise	14%

Of the poor-decision accidents, 108 took place during an emergency. This is when decision making is really put to the test. Some other problem has occurred, and while under extreme pressure the pilot must make cool, calculated, knowledgeable decisions. On 108 occasions the pilot decisions made did not prevent the accident.

Most of the faulty decision-related accidents (78%) took place during the daylight. It was dark during 17% of the accidents, and it was classified as either dawn or dusk for the other 5% of these accidents.

The weather has always had a large impact on pilots' awareness and the resulting decisions that they make. As mentioned earlier, many faulty decision accidents took place in instrument weather conditions. But IFR comes in many forms. What type of IFR conditions produced the most poor-decision accidents?

The following table outlines the visibility that was present at the time of these accidents.

Visibility (sm)

0–3	15%
4–6	8%
7–10	35%
11–20	23%
21+	19%

From this we discover that the majority of the accidents did not take place in IFR conditions. The dividing line between what is IFR and what is VFR is 3 statute miles in controlled airspace, meaning that 85% of these accidents took place in VFR conditions.

When the visibility did constitute IFR conditions (less than 3 statute miles), fog was the obstruction to vision present in the greatest number of accidents (73%). Haze was the next most frequent visibility problem, present in 21% of the IFR accidents, and snow limiting visibility was present in 7% of these accidents.

Thirteen percent of the total faulty-decision accidents took place when it was raining. Some of these were IFR; some were VFR. Another 5% took place in freezing rain or ice.

A low cloud base can also produce IFR conditions and hazardous flying conditions. The following is a table displaying the current cloud ceiling at the time of these poor-decision accidents:

Cloud Heights (ft)

0–50	0.2%
51–100	1%
101–200	1%
201–300	2%
301–400	2%
401–500	2%
501–1000	6%
1001–2000	15%
2001–3000	18%
3001–4000	15%
4001–10000	38%

The table shows us again that more accidents take place when the ceiling is high and within VFR guidelines. There are more VFR accidents simply because there are more VFR flights, but most pilots think of VFR as safer than IFR. These numbers help us guard against having a false sense of security when we fly VFR.

Wind presents as much a problem to pilots as IFR conditions do. In fact, you could argue that wind is a greater threat. If the clouds and the visibility are low, it is often easy to decide against flying. But when there are no clouds in the sky and the visibility is 50 miles, it's hard to say no even when the wind is strong and gusty. What winds offer the greatest difficulty? The following table outlines the wind conditions that were present during decision accidents:

Wind Speed (kts)

0–5	26%
6–10	44%
11–15	17%
16–20	8%
21–25	2%
26+	3%

The wind was not always the cause of the accidents listed here, and probably the reason the speed range from 6 to 10 knots has had the greatest number of accidents is because those speeds are present very often. But what the numbers do tell us is that wind can be a problem even when it is not excessively strong. Wind can be light to moderate and still complicate a takeoff or landing if the wind is across the runway or if the pilot mistakenly selects a downwind runway.

These numbers are only important if pilots use them as a guide on their next flights. The numbers only tell us where pilots have had problems in the past. It appeared that each and every one of these accidents could have been avoided if the pilot had been better able to see what was coming. Being able to project the flight ahead of the actual location of the airplane is, of course, tough, but it can be accomplished with complete situational awareness. When we know where pilots have gotten into trouble in the past, we can raise our awareness when we follow in their footsteps. Because we know better what can happen when we are aware, there will be some times when we decide not to follow in their footsteps, and accidents will be avoided. Awareness has the ability to prevent the next accident from ever happening.

7

Prepare to Be Aware

The accident numbers and ASRS reports seem to cluster around particular areas of operation. This fact should serve as a warning sign. If pilots are repeatedly having problems in predictable situations, then whenever you find yourself headed into a similar situation, you should know to get ready. What if, while driving to work, you saw an accident at a particular intersection four days in a row. Wouldn't you be especially careful as you went through that intersection on the fifth day? Of course you would. You would have heightened awareness to the possibility of a dangerous situation as you got closer to the spot where so many accidents had taken place. There may have been other places along the way where attentive driving was also necessary, but the fact remains: More accidents take place at that intersection than anywhere else that you pass. There are situations that we face in flying that are the same. There are predictable circumstances where we know problems routinely occur. As we find ourselves in these circumstances, we must be ready to raise our awareness. We must prepare to be aware.

The five big accident and incident clusters pertaining to operations among general-aviation pilots are:

1. Traffic pattern operations.

2. Penetration of airspace.

3. VFR flight into IFR conditions.

4. Runway incursions.

5. Overreliance on technology.

The following are examples from pilots who describe in their own words when they found themselves in these situations.

The Classic Uncontrolled Airport Traffic Pattern Conflict

ASRS NUMBER 449200   I had just picked up a newly purchased Cessna 182 for my company. Myself and a co-worker checked the aircraft over thoroughly prior to my taxi, I made all the appropriate uncontrolled airport radio calls. I taxied to runway 2 and did my run-up. During that time, one aircraft entered the traffic pattern for runway 20 and landed. Another aircraft had called inbound but had not yet entered the pattern. I pulled out on runway 2 and began my takeoff roll. Again I announced my takeoff. When I had reached my rotation point the inbound Cessna called entry to the left downwind for runway 20. I proceeded to lift off and climb out from runway 2, making a left turn out of the pattern and away from the Cessna. By this time the Cessna was turning base. The pilot of the Cessna began screaming about what he perceived to be an unsafe situation, and wanted to know why I wasn't making radio calls. The screaming Cessna pilot was very upset and effectively blocked the unicom frequency for a minute or so until someone else on the unicom frequency asked him if he would cease and continue his questions after he had landed. He complied. After a few minutes with the Cessna 182, I decided the aircraft was functioning OK, and went to pick up my IFR clearance home, only to find the #2 radio was not transmitting. So I guess the screaming pilot had a point, I was not getting out of the aircraft with the radio I had selected. However, I must say that at no time was I even close to the arriving Cessna, but did have a hand in upsetting the excitable pilot to the point where he effectively shut down communications at that airport and any others with the same unicom frequency.

At an airport with a control tower, the active runway is determined by the tower controller. But at every uncontrolled airport, the runway selection is a group decision. The decision is easy when a strong wind clearly favors a particular runway, but that is not always the case. When the wind is light or calm, then more than one runway could become active. Some airports have designated a "calm wind" runway to be used when the wind favors no particular runway. The runway selected to be the "calm wind" runway is usually the runway with the greatest clear zone, or the one that directs traffic away from residential areas. But the calm wind runway is not an "official" designation. It is usually something that local pilots have established among themselves and the airport manager. Incoming pilots who do not regularly fly to the airport will not know about the calm runway. When the wind is light or calm, pilots will select a runway that is best for their direction of departure. If headed south, a pilot will want to take off on runway 18 rather than 36. All this taken together means that there will always be differences of opinion when it comes to what should be the active runway. This occasionally leads to conflicts.

When operating at an uncontrolled field, listen for other pilots. A pilot giving a position report such as, "Hometown traffic, 34A is left downwind to runway 24" will quickly tell you everything you need to know. You know which runway is in use. You know the direction of turns in the pattern, and you know the position of traffic in the pattern. If there are no other pilots on the radio, look throughout the pattern for the possibility of nonradio traffic in the pattern. When approaching the airport, if no radio conversations are taking place, that will give you the information that you need, then call the unicom for an "airport advisory." When departing an uncontrolled field, ask if there is a designated calm wind runway at that airport.

Whenever there is a conflict in the pattern, it is always better not to get into a shouting match over the unicom frequency. In the previous example, the "screaming pilot" may have been correct, but he made the situation worse by carrying the discussion out over the radio. Blocking the radio can only create more problems.

Near Midair Collision during Traffic Pattern Entry

ASRS NUMBER 449200   I made an approach from the northeast to Santa Paula Airport, made the appropriate announcements, and approached on the upwind leg at 2000 feet MSL. The upwind was extended for one mile to the southwest of the runway while I listened to an airplane approaching on an extended left downwind from the southwest. The two of us acknowledged one another and I said that I would follow him (and we both announced every position in the pattern). I turned base as the other airplane was almost over the numbers and as I banked from base to final, I saw a large twin flying also on final only 100 to 150 feet under my right wing. Suddenly he (the pilot of the twin) turned left and started to climb. My corrective action was to climb and turn back to the right. Later it was discovered that he never saw my airplane, but was reacting to the airplane on the runway. Instead of doing a go-around straight ahead, he did a 180-degree turn to the left. There were never any announcements from the twin. After landing and watching the twin do yet another 180-degree turn to reenter the final approach and land, I went over to speak with him and find out what he was doing! He said that he had heard nothing on the radio and had made a straight-in approach. He said that he had never been to that airport and did not know the procedure. I am a radio technician and I offered to look at his radios. There was nothing apparently wrong with them. He probably was not very familiar with their operation and the audio panel operation. The flight instructor who was riding with me took the opportunity to remind him of the regulation that says that pilots must familiarize themselves with all information pertaining to the flight before takeoff.

The proper traffic pattern entry will always be a point of debate among pilots. Should you always enter on a 45-degree angle to the downwind leg or are there times when a straight-in approach is allowed? To get to the 45-degree entry position from the other side of the airport, do you fly over the traffic pattern and then teardrop back around, or do you fly around the airport to the 45-degree position? What about making a midfield crosswind at pattern altitude and then immediately turning downwind? The answer is: It depends. It depends on other traffic, the surrounding airspace, and the local practices. There is no set "best" way that will guarantee complete safety in every circumstance. Since there is no guarantee, pilots must prepare to be aware anytime they are near an airport.

Near Midair Collision in the Traffic Pattern

ASRS NUMBER 449200   This incident took place while on a training flight with a foreign student. I had just departed runway

26 but prior to turning left onto a crosswind leg, another aircraft came into view off the right-hand side of our airplane going in the opposite direction. Apparently, what in my estimation was a very close, tight downwind for runway 26. We had to snap left to lessen the closure rate. The cause, in my opinion, was a lack of correct radio usage together with a traffic pattern that did not comply with the AIM and applicable regulations.

The problem is that there are no "applicable regulations." If this near-miss was the result of a "tight downwind" by the oncoming airplane, the question gets even fuzzier. The distance that the downwind leg should be flown parallel to the runway is never stated. I use one-quarter to one-half mile, but that is not a regulation.

Near Midair Collision Due to Distraction in the Traffic Pattern

ASRS NUMBER 445679   I was with a student doing touch and goes in closed traffic. On the downwind of our third touch and go, the tower asked us to fall in behind the Cessna off our right front. I looked for traffic but he was initially blocked by the right windshield pillar of our aircraft. I moved my head to look around the pillar and saw an aircraft (Cessna 172) entering the 45 degree midfield left downwind on a collision course with us. I rolled sharply to the right and descended to avoid him and then I turned back to the left to follow him. All of this took roughly 10 seconds. Contributing factors were: (1) My concentration on my student's procedures and the landing checklist, and (2) I was scanning the base leg and final for the traffic pointed out by the tower and this distracted me from scanning for traffic that might be entering on the downwind. The other aircraft pilot may have been similarly distracted. The tower was very busy and might have been distracted also. However, when he did point out the traffic he said nothing of the unusually close proximity. In my opinion, the tower allowed himself to become overwhelmed. Maybe there were too many aircraft in close traffic, or maybe he just squeezed everyone in too close. Either way both aircraft failed to see and avoid and the tower failed to safely space traffic in the pattern. In high traffic areas and during times of increased workload in the cockpit, it is easy to become distracted and get behind. This goes equally for tower controllers who sometimes bite off more than they can chew. Pilots must maintain their scan for traffic and not get lost inside the cockpit.

Of course it is true that a controller can get overwhelmed, but every pilot should understand that while in VFR conditions they and they alone are responsible for their own separation. If controllers "squeeze" you in too tight for your comfort level, tell them that it is too close and that you need to extend a downwind, or even exit and reenter the pattern.

Many traffic pattern conflicts take place at uncontrolled fields, and we have even read where pilots have suggested that placing a control tower at a busy airport would resolve traffic conflicts. But this conflict (and others to follow) took place at an airport with an operating control tower. So the presence or absence of a controller does not remove the need for pilots to remain aware.

Near Midair Collision during Multiengine Flight Training

ASRS NUMBER 449911 I was conducting touch-and-go flight instruction at Prescott in a Beech Duchess (twin engine). I made one pattern a simulated single engine approach. I advised the tower that we were simulated single engine. Tower acknowledged this while we were right downwind for runway 21R and said, "Roger, traffic to follow on right base to final." I acknowledged that I had that traffic in sight. The controller then said, "Cleared touch-and-go runway 21L. I acknowledged by repeating, "Cleared for touch-and-go runway 21L." This was a change in runways, but at this time I believed that the controller was trying to help us out by giving us the longer runway during our single engine approach. This practice of having students land on runway 21L while being on tower frequency 128.75 (normally 125.3) has happened numerous times before, so we never thought twice about being cleared touch-and-go on runway 21L even though we were flying a downwind for 21R. Turning inbound to runway 21L a near collision occurred between our airplane and another. The pilot of the other aircraft said that we came within 100 to 200 feet of them. We never saw them once.

Expectations reduced this flight instructor's awareness. The controller made a mistake. He told the Duchess to land on the wrong parallel runway, but since the practice of switching runways had happened "numerous times before," the pilot "never thought twice" about the switch. I guess that as a general rule, awareness can only be maintained by always thinking twice. Don't fall into a routine. Every flight, every approach, every landing is a new landing that you have never accomplished before. Treat every situation as unique and you will do a better job of remaining aware.

Traffic Pattern Conflict Results in a Gear-up Landing

ASRS NUMBER 450266   I was approaching Washington-Wilkes airport from the southwest. The wind was calm and there was no other traffic reported, so I elected to land on runway 13. As I lined up with the runway and began my prelanding checklist, I was totally surprised when I met an aircraft leaving from runway 31. There had been no indication of other traffic in the vicinity. I had called in and reported my position and intentions, but the other aircraft was either not using a radio or was using the wrong frequency. The FBO said he was unaware of the other airplane's presence. The encounter upset me and diverted my attention. I simply failed to get the landing gear down. I touched down smoothly and skidded for perhaps 75 feet with no injury to myself and seemingly minor damage to the airplane. No other person was aboard.

Preparing to be aware means allowing time to become aware. When the pilot "lined up with the runway" but had not yet started the prelanding checklist, he was behind the airplane. Each time I roll out on my final approach to a runway and I see the runway numbers, I use that as my last reminder to verify that the landing gear is down. From that point on, I know that I will be busy with the actual touchdown, so I want all other "housekeeping" chores already accomplished. When jobs are not completed on time and therefore left until later, the workload will go up. This makes it harder to have time to watch for other traffic. That's why it would be easy to become "totally surprised" by other traffic. When time is not used wisely and workload gets piled up, any distraction will have a multiplying effect. This pilot became "upset…and…diverted" and then could not get everything done. The distraction created the situation where the landing gear was forgotten.

Two Runways in Use

ASRS NUMBER 457614 As we approached St. Augustine, we called unicom and were advised that they were using runway 13. We entered and called left downwind. A Bonanza called five miles north and he was also advised that traffic was using runway 13. The Bonanza pilot then entered a left downwind for runway 6. We told him that we were downwind for runway 13. Then we saw him just as our TCAS told us to climb. He was at 1400 feet. Later we talked to the FBO person on duty and she stated that the "locals" never paid attention to their traffic advisories. Using runway 6 and 13 at the same time is a dangerous practice. When we were ready for takeoff on runway 13, we could not see the takeoff area of runway 6. Building blocks the view. This is a training base which creates a lot of traffic. Runway 6 is too short for jets. There used to be a control tower there and it needs to be remanned before there is a collision.

Since the use of a particular runway is a group decision, pilots should cooperate with each other. The jet had a reason for using the longer runway 13. The other airplane may have had just as legitimate a reason for using runway 6, but they did not communicate or coordinate, so a conflict came about. The reporter makes a plea to reman a control tower, but as we have read, a tower does not always solve the problems.

Two Airplanes Take Off toward Each Other

ASRS NUMBER 459588   I was testing a new STC for the installation of brakes. The testing parameters required that the aircraft be flown down an ILS until 50 feet above the runway threshold, then land and get stopped within a specific distance. San Marcos is an uncontrolled airport and I broadcast all airplane movements, both on the ground and in the air on the CTAF of 123.05. Prior to start we had calculated and came to the conclusion that we would use runway 26 for takeoff. I taxied down to runway 35 en route to runway 26 and stated my actions on CTAF. I stopped short of runway 26 and performed all necessary checks. I taxied into position and stated my takeoff and intentions again on the CTAF. A visual check indicated a clear runway, so I started my takeoff roll. Full power was immediately used as the aircraft was near maximum takeoff weight. At approximately 100 knots and 2000 feet into my takeoff roll, I saw another aircraft at the opposite end of the runway, rapidly increasing in size. I immediately pulled the engines into full reverse and applied maximum braking. I was moving, slowing, and nearly to a full stop when the other aircraft flew overhead at very close range. I contacted the pilot of the departing aircraft and asked why he did not broadcast his intentions. He replied that he had broadcast his intentions and asked me why I had not broadcasted mine, as he did not hear my transmissions. I stated that I did and that I had witnesses to that fact. I was properly set up and broadcasting on the frequency. I had a conversation with another person who had been listening on a handheld receiver. He stated that my broadcasts had broken up as I approached runway 35 during taxi and that after that point he heard no further transmissions from me. Interestingly enough, radio test personnel who had been near midfield stated that they could hear only my transmissions and never heard the other aircraft. One other contributing factor is that there is a small rise in middle of the airport that obscures the view of the opposite end of the runway. That is why I was unable to see the other aircraft until I was farther down the runway. The rise might also affect the ability to transmit and/or receive the CTAF while on the ground.

Seeing another airplane coming toward you and "rapidly increasing in size" would be a scary sight. Having two airplanes taking off while aimed at each other sounds impossible, but this story points out how it could easily happen. The runway was not flat, and the end of one runway could not be seen from the other. The hump in the runway blocked the pilot's vision, but it may have also blocked the line-of-sight radio transmission. I will remember this story the next time I face a runway that I cannot see over.

Near Midair Collision: Traffic Pattern versus Instrument Approach

ASRS NUMBER 445861   I was conducting touch and go landings in an established right-closed traffic pattern for runway 30R.

A Gulfstream 4 was flying the VOR-A approach at the same time. After making a touch and go, we were flying the upwind leg of the pattern when the Gulfstream passed us at our 9:30 position. We were completely unaware of his position until he passed dangerously close. There was no time for us to take evasive action. I believe that the careless operations of the Gulfstream pilot was a major factor in this near midair collision.

This scenario is played out every day where a standard traffic pattern and an instrument approach terminate at the same runway. In the story the Gulfstream was flying a VOR-A approach, so they were not exactly straight-in, but the problem is classic. Instrument approaches bring the aircraft into the pattern at or near what would be the final approach. Entering the traffic pattern on final or "breaking" the pattern to land straight-in will disrupt the flow of the pattern traffic and create a collision threat. So who is right? The approach traffic or the pattern traffic? Again, it depends. For pilots approaching on instrument approaches, every effort should be made to switch from the approach control frequency to the local "advisory" frequency as soon as possible. Sometimes this means that pilots must remind or even inform ATC that they are switching over to unicom. For pilots in the pattern, you must be aware of where instrument approaches are located. An instrument approach is like a funnel of traffic, so you should always be looking in the direction of where that funnel comes out. Then pilots must cooperate and coordinate. If I am in the pattern and hear an inbound instrument approach, I will extend a downwind to let them in. But the instrument approach pilot should be equally prepared to break off the approach and then blend into the traffic by making a standard pattern entry.

Communication problems are the root cause of the next set of traffic pattern conflicts. Once again, there is no set method for determining which radio frequency should be established as the standard frequency. Airports that have part-time control towers often have a tower frequency plus a unicom frequency. Airports with a Flight Service Station often will also have two frequencies: the "radio" frequency to the FSS and another unicom frequency. To cut down on the confusion, a Common Traffic Advisory Frequency or CTAF system was adopted. This designates which frequency is to be used for uncontrolled field traffic advisories. But confusion still exists when these frequencies are changed or pilots are otherwise unable to communicate their position.

Near Midair Collision When the CTAF Changed

ASRS NUMBER 452670   The Chicago sectional lists an incorrect CTAF for airport C59. I was on a Sunday afternoon pleasure flight, from GYY to C59 and back to GYY. The weather was clear and visibility unlimited. Before departure from GYY, I got a telephone weather briefing from Kankakee FSS. The briefer said he had no notams for C59, but a "note in his file" said that the CTAF at C59 was 122.9. The sectional chart says 122.7. The navigation database in my Loran says 122.9. Approaching C59, I used 122.9 to request "winds and active." I received no response. I flew over the airport at 1500 feet AGL. I saw no activity and no indication of an active runway. Based on a nearby smoke drift, my ground speed, and crab angle, I estimated winds of 300 degrees at 5 knots. This was also consistent with nearby ATIS and another airport landing on runway 36. I flew two touch-and-goes on runway 36 using standard traffic pattern and self-announcing on 122.9. As I departed runway 36 after my second touch-and-go, a Beechcraft Bonanza was approaching to land in the opposite direction. We each veered to our respective rights, missing by 100 to 200 feet. After climbing and leveling off, I switched to

122.7 and sure enough the Bonanza was self-announcing on that frequency. At my suggestion, he switched to 122.9. Later, I heard other traffic using 122.9 for CTAF at C59. I suspect the Bonanza was flying a straight-in NDB practice approach to runway 18. He did not fly a standard traffic pattern.

The unicom frequency is a party line, where everyone gets into the conversation. But when everyone is not on the same channel, conflicts are sure to result. Then remember that radios are not required at uncontrolled airports, so not hearing someone on the radio does not mean that no one is there.

The traffic pattern versus instrument approach problem came into play here as well. This time an opposite-direction IFR approach was in use. Many smaller, uncontrolled airports will have only a single instrument approach, so IFR pilots or IFR students have no choice but to approach the airport from that certain direction. There will be days when the wind favors the runway that opposes the approach. Extra vigilance is needed by both pattern and approach pilots in these situations because the departure path of the runway and the culmination point of the approach are at the same location. Both pilots can be following standard procedures, but they are headed right for each other. When you fly at an uncontrolled airport, you must find out where the instrument approaches are to that same airport, and always expect traffic to appear from the approach.

New CTAF Contributes to a Near Midair Collision in the Traffic Pattern

ASRS NUMBER 457034   A Cessna 172 was involved in a near midair collision with another Cessna 172 during a training flight. During an approach to runway 28 (winds were calm),

I had my student initiate a go-around as part of the training syllabus requirements for that particular lesson. During the upwind climb, my attention was inside the cockpit teaching my student to properly execute the climb while "cleaning up" the aircraft. During our crosswind turn, I looked back towards the runway only to see a second Cessna 172 approach to land on runway 10. Although not 100% certain, it was my belief that this second Cessna had entered the run-way 10 pattern straight-in from the west. It is my estimation that this aircraft passed only 50 to 100 feet below us with a horizontal separation of approximately 300 to 400 feet. It is also my belief that had we not been practicing go-around procedures, we would have collided with the other aircraft either on the runway or during our upwind climb. I believe that this occurred because this airport has only recently been open to the public, and the CTAF frequency at the time was unpublished. Prior to our departure we were unable to find the CTAF frequency in either the Airport/Facility directory or on the sectional chart. I then contacted a nearby FSS and was told that the airport's CTAF was 122.8. When this incident occurred, we were making all the appropriate radio calls on this frequency. The unknown Cessna however made no calls or was transmitting on another frequency. In either case such a pilot is a threat to any and all other aircraft sharing the same airspace.

Of course, the pilot that was accused of being "a threat" probably thought the same thing about the pilot who wrote this story. Anytime there is confusion about the CTAF, you must expect traffic to be present that you have not heard from.

Once again the "problem" of calm winds comes up. Ordinarily, a calm wind is thought of as a good thing as opposed to a strong and turbulent wind. But a calm wind makes the runway selection a coin toss. Your land-ings and takeoffs may be less problematic with calm winds, but your traffic pattern awareness must be increased.

Airport Name Change Contributes to a Near Midair Collision

ASRS NUMBER 455360   The problem arose when I was turning downwind to base for runway 12 at the Avra Valley Airport. Immediately after my turn another pilot called base to final for runway 12 at Marana Airport. I made my turn to final and announced the turn. Then I heard a Cessna say that they were making a go-around. That was when I realized that I had cut him off. Departing the airport an hour later, I overheard two pilots talking on the radio. One told the other that the Avra Valley airport was now being called the Marana Northeast Regional airport and that the Marana Airport was now being called the Pinal Airport. I looked at my brand-new sectional chart. Marana had been properly changed to Pinal, but Avra Valley had not been changed. There had been no change to the Airport/Facilities directory either.

Remember that not only is the unicom frequency a party-line channel for all pilots at one airport, but also the frequency is usually shared with many airports. That is why it is vital that the airport name is mentioned when you give position and traffic pattern reports. When too many airports in the same area share the frequency, it is common for one or more airports to change to a different frequency. You should discuss this with your airport manager if you feel that your unicom frequency has become too congested. The problem is further complicated when airports that have the same runway numbers share a unicom frequency.

In this example, everyone was giving the airport name in their position reports, but it was the airport name that had changed. When the base-leg pilot reported at what he thought was Avra Valley and heard another pilot on final at Marana, he perceived no conflict because the traffic was reporting at a completely different airport. The near-miss took place when both pilots discovered that in fact they were calling the same airport by different names.

Aircraft Lands at the Wrong Airport

ASRS NUMBER 451009   Our scheduled destination was GTR. Approaching the IGB VOR station, my captain called "field in sight" and I informed the controller. The controller then cleared us for the visual approach. We flew a normal approach, announcing our presence on the CTAF and clicking up the pilot controlled runway lights. The VASI appeared in the proper location. We landed on runway 18, which was our intended runway, but it was the wrong airport. We were instead at Lowndes County Airport. The landing rollout, and taxi were uneventful. The captain informed the passengers of our mistake, and called the company who arranged transportation. I canceled IFR. Contributing factors in the situation include five airports within a 15 miles radius, three airports that have the same CTAF, and the common design of a single runway 18/36.

The story is somewhat similar to the story of the airliner that landed in Brussels, not Frankfurt. It seems impossible that a pilot or crew could make such a mistake, but it does happen and again points to the need for awareness around airports. Anytime you are not where you think you are, there is danger. When this airplane landed in the dark at the wrong airport, they were landing on a runway of unknown length and width. They were flying in over unknown obstructions. And they were flying in without the proper frequencies set for CTAF. A pilot in the traffic pattern would not have been able to communicate with the approaching aircraft. The approaching aircraft would not have made position reports for Lowndes County Airport because they thought they were somewhere else, so the aircraft would have "come from nowhere" to another pilot in the pattern.

This class of incidents all took place in and around a traffic pattern. They clearly show that pilots must be at their highest sense of alert and awareness when operating near an airport. Communication is important, but pilots rely too heavily on it. You must fly the traffic pattern with the expectation that airplanes will appear with or without warning. Calm winds actually pose a threat to safety because they do not designate a clear, active runway. Know where the instrument approaches of the airport lead, and then watch for traffic as if nobody else has a radio.

Penetration of Airspace

The lack of position awareness takes place more often than we want to admit. Usually, being "lost" is a temporary situation. We refocus our attention to the problem, and usually we quickly regain a sense of where we are. But occasionally this state of being temporarily misplaced leads pilots into airspace where they have not first gained a clearance. This points out another circumstance where pilots must increase their awareness.

Pilot flies through a temporary restricted area over a flood

ASRS NUMBER 449339   I was hired to fly a photographer on a mission over flood-stricken Franklin, Virginia and Roanoke Rapids, North Carolina. A "self-brief" was conducted with a commercial weather terminal and everything looked fine for the proposed flight. The flight was completed and on my way back to Norfolk, I heard another company pilot request a clearance that was denied due to a temporary flight restriction. That got me thinking. I used ATC flight following from Norfolk to Franklin, terminated and continued to Roanoke Rapids for more pictures. On the way back I contacted Washington Center while within the restricted area, but was told to go back to squawking VFR and to contact Norfolk approach. Nothing was mentioned of the temporary flight restriction. I checked DUATS when I returned and discovered that I had been in the temporary restricted area for some time during the flight. I will never rely on briefing terminals for up-to-date FDC notams again without backup from FSS or a DUAT brief.

The pilot's "self-brief" was not as comprehensive as it should have been, since it did not turn up the existence of the temporary restricted area. He might have been suspicious of the restricted area's existence, since he himself was flying over the flooded area. Temporary Restricted areas are often thrown over areas of natural disaster so that "sightseers" in aircraft will not create a hazard with rescue operations. The DUAT terminals are great, but this story plays to the greatest fears that people have about the use of DUAT. When you speak to a weather briefer, the briefer often knows about information that is pertinent to a flight, but the pilot does not know to ask. The briefer uses common sense and passes along information even though the pilot does not specifically ask. The DUAT computer will only give information that is specifically asked for. In this case, the pilot did not think to ask, so he departed without being completely aware of the situation.

Pilot flies through a temporary restricted area over a forest fire

ASRS NUMBER 449340   I was in cruise flight, level at 8500 feet MSL, VFR direct from CCR to Chester, California on radar following with the air route traffic control center. I had been vectored away from Beale Air Force Base for airshow activities. Approximately 10 minutes later, the center called to request how we were receiving the center frequency. I responded, "loud and clear." Then the center advised that we had penetrated a fire fighting restricted area set up along the 060 degree radial of the Chico VOR at 30 miles with a 13 mile radius. We established our position and advised the center that I was making a left turn to clear the restricted area, which we did at once. The flight was completed without further event. No conflicts with other aircraft was observed or heard on the center frequency. We were unaware of the restricted area and entered it unintentionally. This could have been avoided by my obtaining a preflight briefing (which I did not because of VFR conditions and familiarity with the route of flight) or timely advice from the center controller.

This pilot received no preflight briefing at all. He had fallen into the trap of routine. He was familiar with the route of flight and the weather was good. We should remember that a good preflight briefing goes beyond a weather report. This pilot also relied on the controllers to advise him of the problem. Controllers do a good job of this kind of advising, but it is classified as "additional service" that the controller is not required to give. Ultimately, pilots are responsible for knowing where they are and where they are going.

Unauthorized penetration of Class B after bird strike

ASRS NUMBER 451968   A student and I were on a night cross-country flight. A major objective of the lesson was training within a Class B airspace. We entered from Leesburg, Virginia to Stevensville/Bay Bridge, Maryland. I contacted Dullas approach for a clearance. They "handed off" to DCA approach, but DCA approach would not acknowledge our transmissions at all. We tried repeatedly to speak with them, but received not one acknowledgment. We were forced to navigate around the Class B between Baltimore/Washington (BWI) and Andrews Air Force Base, while transitioning the VFR flyway. Then we suffered a bird strike which covered my side of the windshield in bird parts. Then, while trying to re-establish visual navigation, we inadvertently entered Class B at Baltimore/Washington. I recommended that ATC proceed with the "Washington Tracon" so that once service is established, some continuity can be maintained.

A bird strike at night would be a very scary situation. You probably would not see the bird coming, so you would have no warning that it was a bird. You would not know in the first moments what had happened. There have been instances where the bird came through the window and injured the pilot. Then another problem would be realized. Airplanes don't fly well without a front window to deflect the relative wind. It would be doubtful that a small airplane without a front windscreen could remain in the air. The bird in the story did not break the windscreen, but this pilot really had his hands full. Instructors and examiners can dream up some wild distraction scenarios, but nothing would prepare you for this.

This situation was further complicated with "handoff" problems, as discussed in previous chapters. As pilots approach busy terminal areas, they must prepare for problems and expect the unexpected—even a window full of "bird parts."

Unauthorized penetration of a Class B airspace

ASRS NUMBER 450039 I took off under VFR from Burke Lakefront Airport, runway 24L. Tower cleared me for a right turnout after takeoff because I advised I was VFR westbound. The floor of the Cleveland Class B airspace in that area is 1900 feet MSL and intended to level off at 1800 feet. Using a hand-held VFR GPS, I planned to fly around the inner core until I was southwest of Cleveland and could proceed direct to LUK airport. As reported and forecasted, there was turbulence in the low altitude and I discovered that I had climbed above 1900 feet floor and was at 2100 feet MSL. I believe it was for a very brief period—probably considerably less than one minute—and I immediately descended back to 1800 feet. I think the following contributed to this event: (1) Fatigue—I had been in conferences for three days and had slept poorly for several nights. (2) Insufficient attention–situational awareness in a critical phase of flight, (3) Complexity of the airspace system, (4) Turbulent weather and terminal activity, (5) Lack of effective trim (pitch trim) technique. I did not ensure that the airplane was properly trimmed for level flight, (6) Being in too much of a hurry in the before take-off check. I should have "rebriefed" myself on climb and heading restrictions or should have contacted Cleveland departure control for a clearance through the airspace. As an FAA Safety Program Manager, I have taught and counseled pilots on the use of and respect for the airspace. I'm dismayed and embarrassed that "it happened" to me. Maybe I'll be more effective!

Pilots tend to think that problems will happen to the "other guy." This is a great example of the fact that it can happen to anyone, no matter their experience level. At any time, awareness can insidiously slip away, and any pilot can get caught off guard. The pilot in this story was an FAA Safety Program Manager who routinely gave talks on airspace. Even he became unaware and found himself in a dangerous situation. His analysis of his own mistakes was a laundry list of common awareness traps: fatigue, inattention in a complex environment, being in a hurry. He summed it up best when he admitted that he had lost "situational awareness in a critical phase of flight."

The creeping loss of situational awareness can happen to anyone. Pilots must be ready to reevaluate their situations to uncover problems, but they must also anticipate problems. We must prepare to be aware.

Law enforcement pilot penetrates Class C airspace without clearance

ASRS NUMBER 452410 While flying a surveillance mission, I entered the BUR Class C airspace without radio contact with the BUR tower. I had been flying inbound over Interstate Highway 5 at 1800 feet, talking to Southern California approach control, and awaiting a Class B clearance. As I came close to the Hollywood Hills area, leaving the Class B area, I canceled my clearance request and my radar services were terminated. I was about 1 mile from the BUR surface area, and while looking at my chart for their tower frequency, I entered the Class C surface area. I was one and one-half miles inside when I realized my mistake. I am a new pilot with the county law enforcement unit. On this date I was working with ground units engaged in a mobile surveillance. I was monitoring other aircraft in the area, my ground units, and communicating with my observer. I got behind on my cockpit workload. I believe my inexperience played a role in this violation. There was a fatigue factor, due to long hours this week. Our crew consists of two pilots, and I should have let the other crewmember fly while I observed, due to the fatigue factor.

Fatigue in a complex environment. Lack of experience and distractions are all themes that recur throughout the incident reports. As pilots we know these problem areas exist, so we should not get blind-sided. When pilot workload goes up, awareness must follow.

Continued VFR into IFR Conditions

Every year it happens. Pilots press on into deteriorating weather conditions. It is the leading cause of fatal accidents among pilots. But some pilots do survive their encounters with the clouds and live to tell about it.

VFR into IMC with no clearance

ASRS NUMBER 451785   I departed Selma, Alabama en route to South Florida. The weather was 1900 feet broken with 6 miles visibility. As I progressed to 15 nautical miles west of Troy Airport the weather started to lower. Nearby Dothan, Alabama was reporting a 700-foot overcast with 2 miles visibility. As I approached Dothan the weather dropped to 400 feet and 2 miles visibility. The overcast showed breaks above and the tops had been at 2000 near Selma. I was 5 miles east of Carl Folsom Airport when I went into the clouds on a 170 track to climb on top. Due to the cockpit workload in this type of aircraft (Sukhoi-31) I was unable to contact Dothan approach. I maintained about a 160-degree track until I broke out on top of the clouds. I stayed VMC and climbed up to approximately 7000 feet to get around some other clouds. This was approximately 15 miles northwest of Genavo, Alabama. At 15 to 20 miles east of Genavo I was unable to descend through the scattered/broken deck. Due to the fact that I was in an unstable aircraft, I was unable to use the radio and change frequencies to contact Dothan approach control.

This pilot was in danger two ways. First, there was the threat of spatial disorientation. Second, there was the threat of colliding with another aircraft while inside the clouds without a clearance. This pilot did live to tell the story, but did not seem to learn much from the experience. The following three pilots had encounters with instrument conditions, and it seemed to change their life.

VFR pilot flies out over a layer of clouds

ASRS NUMBER 41835   Problems encountered include thickening cloud layers below the altitude I flew, and unfamiliarity with the phrase: "What are your intentions." Preflight showed no adverse weather except for broken cloud cover at southern

Kentucky border for the time of arrival. Few clouds were observed 20 miles northwest of Frankfort, Kentucky, but believed them to be incomplete and would dissipate soon. Initial call-up to Lexington approach disclosed IFR and mist, plan B was to fly to Marshall Field. I was given a transponder ident and a heading to descend on. I expected to find Marshall without delay. However, the clouds were much thicker. I remember three distinct layers on my descent to 1,800 feet and thinking that my normal rate of descent would only prolong my time inside the clouds. I used carb heat, reduced power to keep RPMs below red line and increased a wings level descent to 1,800 feet per minute. I kept repeating to myself that "instruments don't lie" and remembered the hood drills and techniques I had worked on with my instructor. My awareness of the severity of my situation built slowly but now has made a permanent impression about safety issues in my mind. I overshot Marshall and had to be vectored to Cynthiana. Upon visual confirmation of Cynthiana, I signed off and thanked ATC for their help. But, instead of landing at Cynthiana, I switched my mind and opted to travel to Marshall Field by setting on a course west of the Lexington VORTAC. I found Paris (via watertower identification) and followed what I incorrectly identified as the road west to Marshall. Realizing the wrong course, I radioed Lexington Control, requested directions to Georgetown after heading back to Paris. I now more than ever realize the hazards clouds present and the importance of maintaining visual contact with the ground. I never want to go through the clouds again with so little experience. I also plan to take every effort to update my weather information from inflight advisories and metars, and am currently taking additional safety training at our FBO. I am also more determined than ever to start IFR training, hopefully next summer.

Pilot climbs into a fog and becomes disoriented

ASRS NUMBER 418559   I arrived at Batten Field, Racine, Wisconsin. The weather was poor, both there and at my destination—Pontiac, Illinois. I chose not to leave Racine until the weather improved. I continually called flight service, Green Bay, to look for improvement. Weather to the south of Racine was gradually improving, but Racine remained socked in. Later I became anxious to go in order to get back to my business. Unfortunately, I decided to fly despite the fog, figuring I'll climb on top and descend close to home. At that time Pontiac had 1,900 feet ceiling. Upon climbing into the fog, I became disoriented. I immediately called back to Racine to let them know what was happening. I then arrested my problem by "flying the airplane" although my body was telling me I was in a slow rightturning descent. Once I leveled out, I contacted Milwaukee departure and they vectored me through the clouds. My decision to ignore weather in order to get back to my business was, to say the least, irresponsible. I respected the weather, but my business sense told me otherwise. The fact that I lived through this experience brought home just how dangerous weather can be if one chooses to fly in it. My trouble in flight was only a few seconds, but my memory of it will last a lifetime.

VFR pilot presses on into IFR conditions

ASRS NUMBER 420767   I wanted to move my plane from its summer airport to its winter airport. The weather forecast was calling for snow the following weekend and I figured that I would not be able to get the Cessna 150 to its home base if I waited. So with 3,000 feet ceilings, I blundered off into the east skies with lowering ceilings ahead. Five minutes into the flight, I looked at the compass and said to myself that 180 degrees from this is my way out! Well, let me share the angst/terror, etc. I hit the wall of clouds and all of a sudden could not see the ground! I was only five miles away from my destination and thought that I could just follow the road into my home strip! This was the biggest mistake of my life! After 4 or 5 minutes of total IMC, I finally got somewhat of a grip. I firewalled the throttle and climbed back up to 2,400 feet and prayed. I also headed in the direction that was my escape. I totally relied on the attitude indicator for help. For those of you who haven't been there, let me tell you now, that what you have learned about flying IMC is all true! Thoughts of how my attitude indicator sometimes decides to tumble zoomed through my head. A break in the clouds granted me a visual reference for a moment. Then more clouds…light…and a break about 4 miles back to home. I could see where my home base should be. Just when I could almost see home the ceiling enveloped me again and I descended to 2,000 feet MSL. The home base appeared right where it was supposed to be. At 110 mph I turned sharply and the runway was in sight! Slow to 80 mph, full flaps, idle, full rudder slip and I was on the ground! I taxied to my tie-down and said a prayer to my God that I was still alive. I do not write this as an adventure/thriller but to help others not to make my mistake! I could have easily been on the NTSB list of accidents that we read. I truly thought that I could end my life in muck. It is not funny or heroic by any means. I was stupid. I could go on and on about what happened to me but you get the point. Do not scud run or fly into IMC—period!

This pilot says it all—Don't scud run or fly into IMC—period!

Runway Incursion

The lack of awareness continues to be a source of problems when aircraft move on the ground. Not long ago, runway incursion was not even an accident category, but today with more traffic and large airports, it is as easy to get lost on the ground as it is in the air. The rule among pilots must be that heightened awareness must begin from the moment the airplane is untied or pulled from the hangar.

Airplane takes off over the top of another

ASRS NUMBER 460250 There was not much traffic on the ground control frequency, and I asked to taxi for departure.

My position was on the northern FBO ramp. The controller advised me to taxi to runway 35 via the taxiway, onto runway 17, then back taxi to runway 35, and hold short of runway 26, and to contact the tower for further taxi on runway 17. I had been to this airport about a month before and so I glanced at the airport diagram on my commercial chart to identify the taxiway that would start me on my way. I also at this time looked for a parallel taxiway to runway 35, but there is none. As I came up to runway 17, and made the left turn to begin my back taxi, I observed a military aircraft rolling out on a runway (it was runway 3). I made a mental note of the runway I was to hold short of, and I continued my taxi. I then switched to the tower frequency. About the same time that I switched, I looked out my left window and saw a Bonanza right over my head with landing gear in the well. My first thought was: "that's strange, they let an aircraft depart over me while I was taxiing." My next thought was the realization of what I had done. The local controller then called and inquired in a demanding way, "weren't you told to hold short of runway 26?" I admitted to him that I was told that, but that the aircraft downfield had confused me and that I thought that was the runway I was to hold short of (runway 26 in my mind, was actually runway 3). Of course my mind was now a blur. I could not believe what I had just done. The departing Bonanza did not report anything to the tower, so I assume that he rotated about 1500 feet away and did not see me. He crossed over me at about 50 to 75 feet and his departure profile seemed normal. When the tower released him, there was no disturbance in his voice or questions about what had just happened. I think he was totally oblivious to the incident. I was asked to go to a discrete frequency where all my pilot data was collected. I then collected myself and resumed the flight. The Bonanza had departed from my blind spot in the cockpit and I was totally surprised at a threat coming from that area. I seemed to key on the military aircraft that was rolling on runway 3, because in my mind, I fixated on that aircraft and was sure that was where any conflicts would come from. Another contributing factor was the back taxi on the departing runway. It is very rare at a controlled airport to back taxi the full length of the departing runway. Add to this the fact that ground markings are completely different once on a runway. I think it is a matter of conditioning, whenever you cross a runway you see the yellow dashed and solid hold lines and you know you are crossing a runway. This was not the case for me, I saw no yellow lines that would have made me look twice.

Whenever you are not sure of a taxi clearance or are unfamiliar with the surroundings, you can ask for "progressive taxi instructions." This means that the controller will relay your instructions as you proceed. This can cut down on misunderstandings about the location of turns and where you should stop. The pilot's expectations also contributed to the previous problem. He said, "It is very rare at a controlled airport to back taxi the full length of the departing runway." It may be rare, but every flight is different. No pilot should rely on the routine of the past to predict the future.

A pilot's ELT goes off leading to an incursion

ASRS NUMBER 460325   I am a student pilot. I flew to the practice area for maneuvers, and then I headed back to the airport with the intention of practicing touch-and-go landings. I radioed the airport on the control tower frequency of 126.00. I had just finished my descent to pattern altitude on the left downwind for runway 27, when suddenly I encountered strong turbulence. My head slammed up against the ceiling, which knocked my headset partly off. This happened even though I had my shoulder harness and lap belt fastened low and tight. I recovered quickly and returned the airplane to straight and level flight. The turbulence subsided. At that point, I noticed an emergency alarm sounding in my ears. I immediately thought that my airplane's ELT must have been activated by the turbulence. Although the alarm was loud and interfering, I was able to maintain radio communications with the tower. I received landing instructions. Tower communications were busy and I decided not to take time to make any comment concerning the ELT. After touchdown, I exited runway 27 at the very first intersection. I stopped the airplane and completed my after-landing checklist. The ELT alarm was sounding much louder in my ears after I got on the ground, and I could not seem to hear any radio communications. I decided to switch to ground control frequency 121.7, hoping that I could get taxi instructions. The ELT was now even louder in my ears. I made two transmissions attempting to get taxi instructions. I could not hear any coherent responses to my communications other than a low static rumbling and an occasional broken word or two in the background. I decided that the situation was unsafe, and that I should take some action to clear the area. I began to taxi. I then heard some broken words under the ELT alarm sound. I stopped the airplane and transmitted again on 121.7, saying I was unable to copy any voice transmissions. I realize now that I should not have taxied without clearly understanding instructions. I assumed that everyone else was hearing the same loud alarm that I was hearing, and that they understood that my airplane was the source. I looked up at the tower windows for a few moments, and then decided once again that this was an emergency, and that I should clear the area. Even though I had been trained to look for light gun signals in the event of a radio failure, I mistakenly failed to equate this situation with a "radio out" situation. I didn't see any light signals, but I have to confess that I forgot to look for them. I was concerned that my presence at the airport with an active ELT constituted a danger to others. The main thought in my mind is that I should taxi back to the FBO so someone could deactivate the ELT transmitter. I taxied to the hold short line of runway 1/19 and looked for traffic. I saw no traffic, so I taxied across runway 1/19 and continued towards runway 14/32. At that point, an airport safety vehicle suddenly appeared on my left wing tip with flashing lights. I also heard the first clear radio transmission since I had landed with a voice saying clearly above the sound of the ELT, "stop the airplane!" The safety vehicle pulled in front of me, waving with a gesture that said "follow me." I followed the vehicle across runway 14/32. At that point I was directed to the FBO. I parked the airplane and was immediately joined by a mechanic and the airport safety officer. I was asked to call the tower. I was very surprised to be told that I had committed a runway incursion violation. The controller gave me a stern lecture. He said I should have tried again on 126.00, or to have waited for light gun signals, or to have waited for the safety vehicle. I now understand that a small airplane parked on a wide taxiway does not constitute nearly so great a danger as the same small airplane if it taxies without proper clearance.

When in doubt while on the ground at a controlled field—stop. Don't stop on a runway, of course, but it is better to hold your position than to move about unsure of what to do.

Distracted by Automation

Although not reflected by a large number of incident reports yet, there is a growing need for greater awareness when using and learning to use new technologies. When automated systems are given the task of maintaining position or aircraft control, pilots can lose awareness. Pilots give away too much control and rely too heavily on the automated systems. It is true that too much of a good thing can be dangerous.

A pilot inadvertently penetrates Class B airspace using his new "toys"

ASRS NUMBER 449901 I was returning on a multiengine training flight. My student had chosen 6500 feet MSL for an en route altitude. At the time he announced that altitude I knew this would be a problem approximately 15 nautical miles later, but elected to allow the student time to realize that

6500 would be a potential incursion into the Phoenix Class B Airspace. I decided to correct the situation myself, if the student hadn't already, by the time we reached Olberg, Arizona. That would have allowed us time to descend below 6000 feet well before reaching the point where the floor of the Class B lowered to 6000. The aircraft was equipped with a GPS and an autopilot. The student asked me to demonstrate and explain the features of this equipment. This was only the student's second flight in the airplane so I instructed him how to engage the autopilot and then proceeded to explain some of the more often used functions of the GPS. I began programming the vertical navigation, and entered the required rate of descent, when I noticed our position was roughly 3 nautical miles southeast of Sun Lakes, Arizona. The "cue" to begin descent by the GPS had not activated, and we were watching the GPS display but not monitoring our position. When I realized that we were, in fact, inside the Class B airspace I immediately instructed the student to disengage the autopilot. I immediately exited Class B airspace. We have many aircraft of the same make/model/type at our company. But these are the first aircraft purchased that have the GPS and autopilot. We have identified the "problem" already, and have cautioned the instructors and students that extra vigilance is necessary when instructing on, and experimenting with the new "toys."

A flight crew almost strikes an obstruction while on a visual approach.

The pilots flew the "visual" approach with heads down in the cockpit.

ASRS NUMBER 446926   Arriving late at night to Spokane International Airport (GEG), I was anticipating an ILS approach to runway 21 based on the forecast winds. GEG does not have automated ATIS via our company frequency, so I was caught a little off guard when we finally received ATIS from the conventional communications radio and found out that we would be shooting a visual approach to runway 25. I assumed that this would be an uneventful approach using a Precision Approach

Path Indicator (PAPI) for vertical guidance, but in hindsight I was complacent in preparing for a night visual approach into an unfamiliar airport, after a long day and with fatigue setting in (on duty 11 hours, and awake 16 hours with a short sleep the night before). Arriving from the east, we encountered the runway visually while at 7000 feet and approximately 18 miles from the airport. We reported the airport in sight to approach control and they cleared us for a visual approach. I selected 4000 ft and initiated a descent to that altitude. At approximately 12 mi out and descending through 5000 feet, I finally noticed a group of very tall lighted towers ahead and slightly left of us. I leveled off and stated this to the first officer. Approach control then asked, "do you have the towers in sight?" indicating that they were concerned. I then initiated a climb up to 5500 feet and visually cleared the towers by approximately 1000 feet and slightly to the north. Although this was a visual approach, I was still heads down in the cockpit a lot using our glass automation (map display) and vertical path indications to fly in to the runway. The Flight Management System (FMS) vertical path indications showed me to be high, so I was descending to reacquire the FMS path. This could have put me in conflict with these towers. I was using my familiar FMS guidance, when I should have been heads up, using the PAPI, and looking for outside visual cues for this approach. Fortunately I did notice the towers visually and they were slightly south of our course, but a conflict could easily have occurred given my mind-set, unfamiliarity with the terrain around Spokane, and fatigue.

The FMS computer is a wonderful piece of equipment to have. It is designed to process information so that pilots can use their time more efficiently. The system is supposed to present the pilot with the "big picture" as an aid to situational awareness. But in this case the overdependence on the automated computer system actually reduced the pilot's awareness and would have caused an accident if the pilot had not at some point left the newfangled equipment and simply looked out the window.

Index

About the Author

Paul A. Craig, Ed.D., longtime pilot, flight instructor, aviation educator, and author, has designed and conducted extensive research on the high accident rate among general-aviation pilots. For this research, he earned a doctorate in education, with special emphasis on pilot decision making and flight training. A Gold Seal Multiengine Flight Instructor and twice FAA District Flight Instructor of the Year, he has spoken widely to flight instructors and others on improving flight training and safety. He is the author of *Pilot in Charge, Be a Better Pilot*; *Stalls & Spins*; *Multiengine Flying, 2nd Edition*; and *Light Airplane Navigation Essentials*, all from McGraw-Hill's renowned *Practical Flying Series*.